T0357212

Praise for *Live Forever?*

'A beautifully written book about life and death: busts myths, explores the stories of science and history, and ultimately is rich and uplifting.'

Adam Rutherford, author of
A Brief History of Everyone Who Ever Lived

'I heartily recommend this deep dive into wellness and the human body by Professor John Tregoning. It turns out that none of us are getting out of here alive, but John has bravely put his mind and body on the line to discover the secrets of staying healthy and happy as we get older. A funny, fascinating and educative book that has convinced me to get rid of my pigeon coop and renew my gym membership.'

Ben Willbond, co-creator of *Ghosts*

'Dr Tregoning returns with another wonderful book about life, death, the body and pretty much everything else. *Live Forever?* is witty, insightful, detailed without being overwhelming for the non-scientist, and deeply thought-provoking. I read it with bated breath . . . light-hearted, optimistic and often laugh-out-loud funny. Essential reading for anyone who's not immortal.'

Frank Turner

'Warm, uplifting and laugh-out loud funny. But make no mistake, this is also a profound journey through countless frontiers of modern science, jam-packed with big ideas about what it takes to live well and be happy. Ostensibly about death, what comes across is a passion for life.'

Daniel M. Davis, author of *The Beautiful Cure*

'The idea that a book about death and dying could be a delightful read seems ridiculous, and yet John Tregoning has pulled off this trick with humour, flair and elegance.'

Nessa Carey, author of *The Epigenetics Revolution*

'Taking a look under the bonnet of the human body, John Tregoning outlines how far we've come with medicine, and how far there may yet be to go if we are going to *Live Forever*.'

Al Murray

'Utterly brilliant. Tregoning manages to make scientific scepticism both hilarious and life-affirming. Madcap self-experiments and a healthy sprinkling of toilet humour compound to make a wonderful introduction to human biology and the scientific method.'

Dr Monty Lyman, author of *The Immune Mind*

'A thoughtful and highly accessible book from one of the UK's leading bioscientists that presents a look at healthy living from an honest and balanced perspective.'

Sir Jonathan Van-Tam, former Deputy Chief Medical Officer for England

'The starting point for John Tregoning's *Live Forever?* is John wondering about how he might die and whether he can do anything about it. Despite the seriousness of the topic this is an at times hilarious account of all the things we might die from . . . With the media full of advice on how to live a healthy life, and billions being spent on research into how to slow ageing, *Live Forever?* is exactly the book we need to understand our own mortality and what we can do to live a long and healthy life.'

Luke O'Neill, author of *Never Mind the B#ll*cks, Here's the Science*

LIVE FOREVER?

*A Curious Scientist's Guide
To Wellness, Ageing and Death*

PROFESSOR JOHN S.TREGONING

ONEWORLD

A Oneworld Book

First published by Oneworld Publications Ltd in 2025

Copyright © John Tregoning 2025

Illustrations by Ash Uruchurtu

ISBN 978-0-86154-938-2
eISBN 978-0-86154-939-9

Typeset by Hewer Text UK Ltd, Edinburgh
Printed and bound in Great Britain by Clays Ltd, Elcograf S.p.A.

Oneworld Publications Ltd
10 Bloomsbury Street
London WC1B 3SR
England

Stay up to date with the latest books,
special offers, and exclusive content from
Oneworld with our newsletter

Sign up on our website
oneworld-publications.com

For my parents who have already travelled this road,
And my children who are just starting it.

A Cautionary Note

Self-experimentation is not without risk. I have twenty-five years' experience in the lab and work very closely with medical doctors. There is risk in everything – and anything you do is at your own risk.

Contents

And if all you ever do with your life
Is photosynthesise
Then you deserve every hour of these sleepless nights
That you waste wondering when you're gonna die

Frank Turner, 'Photosynthesis'

And since to look at things in bloom
Fifty springs are little room,
About the woodlands I will go
To see the cherry hung with snow.

A. E. Housman, 'Loveliest of Trees', *A Shropshire Lad*

Time and Tide

Do not go gentle into that good night,
Old age should burn and rave at close of day;
Rage, rage against the dying of the light.

> Dylan Thomas, 'Do Not Go
> Gentle into That Good Night'

Time wants you dead. Not just you, but everyone you have ever met. It wants your heart to stop and your lungs to fail. It wants your muscles to spasm and your bones to ache. It wants your mind to dwindle, your brain to wane and you to fade from the world, your only traces residing in the memories of others.

But not *me*, or at least that's what I thought when I was young, because like all of us when in our youth, I was (in my mind) *immortal*.

But then comes a moment when we seriously contemplate our mortality. For me, this fell on 1 December 2022, just after I pulled out my first grey chest hair. I then noticed that the lighter hairs on my head were not, in fact, blond as I had been reassuring myself, but also grey. To compound matters, I'd started moving my mobile further away to read text messages – the inevitable drift into long-sightedness. It felt like a crossroads: entering my 'late forties' and so the second act of my life. I turned to pondering how I might

1

die . . . and if anything could be done about it. Live *forever*? Of course not. Sadly, Time does eventually win. The brutal truth is that we are all going to die, even me. Anyone who tells you otherwise is selling something.

I turned to pondering how I might die . . . and if anything could be done about it. As a scientist, I had the curiosity, expertise and crucially the contacts for a journey of self-discovery. I was not the first person to investigate the causes of death; conveniently, the World Health Organization (WHO) lists seventeen thousand possible disease types (infectious and non-infectious) in the latest International Classification of Diseases (ICD11) guidelines. Many of these diseases are quite rare. And most of them won't kill you: for example, sleep-related bruxism – teeth-grinding in your sleep – is unlikely to prove fatal, unless it prompts your partner to smother you with a pillow.

Making a few assumptions, I can narrow down my march to mortality. Firstly, I will probably live out my three score years and ten. In fact, my current life expectancy as a man living in England is 79.3 years, a whole 2.5 years longer than my Welsh compatriots, though 3.8 years less than my wife. Secondly, living in a high-income country in the twenty-first century makes death from some terrible infectious disease improbable (see *Infectious*[1]). Thirdly, I am unlikely to die of trauma – and even if I were, there isn't much I can do about it.* Therefore, death would come calling because one (or several) of my organs has failed me. We call these types of diseases 'non-communicable' because you can't catch them (except as

---

* I am choosing to ignore Armageddon because getting old is enough to worry about without having to fret about nuclear holocaust (which was drifting in and out of popularity at the time of writing).

we will see later, this isn't exactly true – you can be infected by something that ultimately leads to organ damage).

The vast steps forward made by science in the prevention and treatment of infectious disease have basically doubled average life expectancy in the last century. We are all now much more likely to die of non-infectious diseases than infectious ones. Improvements in infection prevention and their effect on causes of death are borne out in recent stats from the WHO; non-communicable diseases accounted for three-quarters of all lives lost in 2022.[2] Reflecting this global change, the causes of mortality in the UK have changed dramatically over the last century (Figure 1).

Whilst the reduction in death from infection should rightly be celebrated, the huge progress in infectious disease prevention and treatment sends organ failure to the top of my kick-the-bucket list. I therefore wanted to investigate which of the seventy-nine organs in my body might fail. My remaining time feeling short, I didn't want to squander it worrying about all seventeen thousand different diseases catalogued in the ICD11 codes; to prioritise, I consulted the oracle of British life – the Office of National Statistics (ONS). The ONS captures a cornucopia of detail from Aaron's Hill to Zouch, including: a list of deprivation in coastal towns; a breakdown of internet users (surprisingly not just one category of 'everyone'); and of course the tables I wanted, those relating to death.[3] The big five causes of death (excluding pneumonia, a type of infection) in 2022 are shown in Table 1. It admittedly represents a high-income country; infection still kills many people (especially children) in low-income countries. The top causes for men and women fall in a subtly different order and we will explore these gendered differences. But this should be seen as rearranging deckchairs on the *Titanic*; we all sink eventually.

Male

	1915	1925	1935	1945	1955	1965	1975	1985	1995	2005	2015
1-4	I	I	I	I	I	I	H	H	I	I	C
5-9	I	I	I	M	M	M	M	M	C	C	C
10-14	I	I	M	M	M	M	M	M	M	M	C
15-19	I	I	I	I	M	M	M	M	M	M	M
24-29	I	I	I	I	M	M	M	M	M	M	M
30-34	I	I	I	I	I	H	H	M	M	M	M
35-39	I	I	I	I	H	H	H	H	H	M	M
40-44	I	I	I	I	H	H	H	H	H	H	M
45-49	I	I	I	I	H	H	H	H	H	H	H
50-54	I	I	I	I	H	H	H	H	H	H	H
55-59	I	C	C	H	H	H	H	H	H	H	H
60-64	I	C	C	H	H	H	H	H	H	H	C
65-69	I	C	C	H	H	H	H	H	H	H	C
70-74	I	D	C	H	H	H	H	H	H	H	C
75-79	D	D	H	H	H	H	H	H	H	H	H
80+	D	D	H	H	H	H	H	H	I	H	H

Female

	1915	1925	1935	1945	1955	1965	1975	1985	1995	2005	2015
1-4	I	I	I	I	I	I	H	H	I	I	C
5-9	I	I	I	M	M	M	M	M	C	C	C
10-14	I	I	I	C	M	M	M	M	M	M	C
15-19	I	I	I	C	M	M	M	M	M	M	M
24-29	I	I	I	M	M	M	M	M	M	M	M
30-34	I	I	I	I	I	M	M	C	C	M	M
35-39	I	I	I	I	I	C	C	C	C	C	C
40-44	I	I	I	I	C	C	C	C	C	C	C
45-49	H	I	C	C	C	C	C	C	C	C	C
50-54	H	C	C	D	C	C	C	C	C	C	C
55-59	H	D	C	D	H	H	H	C	C	C	C
60-64	I	D	C	D	H	H	H	H	H	C	C
65-69	I	D	C	D	H	H	H	H	H	C	C
70-74	I	D	C	H	H	H	H	H	H	C	C
75-79	D	D	H	H	H	H	H	H	H	H	C
80+	D	D	H	H	H	H	I	H	I	I	D

Key

I	Infection
C	Cancer
M	Misadventure
H	Heart
D	Dementia

Figure 1. Changes in causes over time and age. Indicates the most common cause of death in that age group for that decade. Data from the Office for national statistics. www.ons.gov.uk/peoplepopulation-andcommunity/birthsdeathsandmarriages/deaths/articles/causesof deathover100years/2017-09-18.

MEN			WOMEN		
Cause of Death	Numbers	Proportion of deaths (%)	Cause of Death	Numbers	Proportion of deaths (%)
Ischaemic heart diseases (heart attack)	38,730	13.3	Dementia and Alzheimer's disease	42,635	15.0
Dementia and Alzheimer's disease	23,332	8.0	Ischaemic heart diseases (heart attack)	20,626	7.2
Malignant neoplasm of trachea, bronchus and lung (lung cancer)	14,856	5.1	Cerebrovascular diseases (Stroke)	16,228	5.7
Chronic lower respiratory diseases (particularly COPD)	14,690	5.0	Chronic lower respiratory diseases (particularly COPD)	15,125	5.3
Cerebrovascular diseases (Stroke)	13,046	4.5	Malignant neoplasm of trachea, bronchus and lung (lung cancer)	13,715	4.8

Table 1: Causes of death in the UK. ONS data from 2022.

And whilst the ICD list contained 16,995 too many ways to die, I felt that a list that covers only three organs – the heart, the brain (twice) and the lungs (twice) – didn't fully reflect the capacity of my body to betray me. For completeness, I decided to also worry about cancer in general, diabetes, kidney and liver failure, which account for places 6, 7, 8 and 9 in most lists of non-communicable causes of death. Of course, this list isn't exhaustive, but it does cover the main players.

They say knowledge is power. But just knowing the ways I could meet my maker didn't feel that empowering. Drawing upon my twenty-five years' experience as a research scientist, I decided to look further into these causes of death. I approached my mortality as I would a problem in the lab, and took a reductionist approach to explore how disease affects the normal working of each organ, one by one from a cellular level upwards. In my initial hubris, I assumed that health could be viewed entirely as a physiological thing – if you understand the organs, you understand disease.

My first (not so shocking) observation is that everyone dies. Admittedly it didn't take a quarter century of scientific training to discover this. Famously, we cannot avoid death. The inescapability of death came up again and again as I researched the book. Death is the original zero-sum game;[*] as seen in Figure 1, reducing death from one cause pushes up death in another. Despite the catchy title of the book (and whatever tech billionaires may hope), one cannot, and should not, live forever. And whilst our perceptions of ageing change as we age (Figure 2A), our chances of dying exponentially increase with our years (Figure 2B).

Since we cannot avoid death, can we do anything to delay it or at the very least make the home stretch less unpleasant? There are four basic tenets that hopefully everyone knows already. Eat less, do more, don't drink, don't smoke. But why do the big four contribute to healthier life? Critically, are there other things? A wealth of ideas exists about how to reduce the impact of ageing and the likelihood of

* Zero-sum games are ones where one person wins at the expense of another – Monopoly being the classic example: the gains I accrue when my children land on my Mayfair Hotel are their losses. They often make everyone except the winner quite sad, which is why after calling out 'Rent', I often find myself collecting my money and the board pieces from around the room.

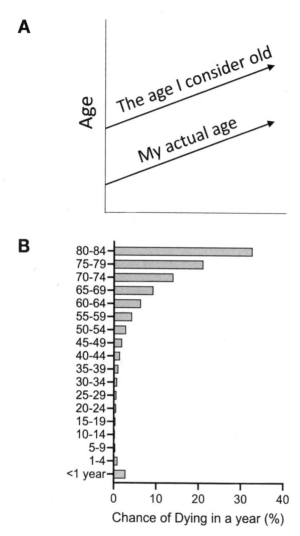

Figure 2: **Ageing and our relationship to it.** A) Ageing is subjective (adapted from an online meme – attribution unknown). B) We all die! Image adapted from OurWorldinData. Data from WHO, Global Health Observatory (2022).

premature death, often lumped together as 'wellness'.* Some of these ideas are scientific, some semi-scientific and some frankly unrooted in any form of science whatsoever, or even common sense for that matter. I drew upon my experimental experience to investigate whether any of the more commonly suggested approaches can reduce the impact of ageing on my organs as I enter the second half of my life, and whether science has made any progress in delaying death (spoiler/not spoiler – of course it has). I am going to share the tools I used to look deeper into the claims, to better arm you against the next sweeping online claim that eating nothing but blueberries will stop you getting dementia.

As an immunologist, I spend most of my days thinking about how our bodies fight off infections. As we will see, immunity is finely balanced; while it keeps us fit and healthy in our prime, an immune system in overdrive or gone haywire can contribute greatly to our decline. Because of my life in science I feel qualified to look at ageing, disease and death, but as a scientist I have questions and uncertainty. Science is not (yet) the whole picture – the underpinning rules by which our bodies work are immutable, but we scientists only glimpse them through a glass, darkly. While science (which is humanity's knowledge of nature) constantly evolves, 'the rules' themselves do not change. There remains so much knowledge yet to be gleaned. One of the challenges is to supress my cynicism (honed by two and a half decades of research) that most anti-ageing tips are bunkum, thereby dismissing an entire field of work. I have, as far as possible, taken an open-minded approach to explore the scientific basis behind 'wellness hacks'. I've tried

* Whatever that means.

to find published evidence from studies enrolling large numbers of people (called epidemiological or population level studies). If the epidemiology supported an idea, I drilled deeper to evaluate its biological plausibility.

If, and it's a big *if*, the intervention proposed passed my plausibility threshold, I tried it on myself. As I explored each organ, I performed experiments upon myself; these got increasingly desperate as I moved through the body, discovering that time was not a tide I could turn. This may sound a bit extreme, but self-experimentation has a long (and sometimes noble) history. Keyhole surgery, used to operate less invasively on hearts, descends directly from Werner Forssmann passing a urinary catheter up one of his own veins and then using an X-ray to show the catheter had reached his heart. Barry Marshall famously (or infamously) proved that an infectious bacteria (*H. pylori*) caused gastric ulcers by drinking said bacteria.

At the other end of the scale of usefulness exists a whole range of daft experiments people performed on themselves. Dr Akira Horiuchi won an Ig Nobel Prize for his work on self-colonoscopy.[4] John Stapp tested how fast a human could accelerate from a stationary position by attaching himself to a rocket sled, reaching 632 mph – and bursting blood vessels in both his eyeballs. And I'd be remiss not to mention Michael Smith, who allowed a bee to sting him daily on different body parts in order to produce a pain index – unsurprisingly, being stung on the shaft of the penis was the most painful.[5] But for sheer idiocy, one would be pushed to beat the American surgeon Nicholas Senn, who pumped six litres of hydrogen up his anus (presumably to make his farts squeakier). My ambitions are much more modest: I want to identify changes I can make as I enter middle age that have a long lasting and beneficial impact on my health.* I'm hoping the experiments help answer some of my questions.

Before I start pumping gas up my bottom in the name of science in my quest to live (better) forever, I need to understand what death and ageing actually are.

---

* A note on self-experimentation. For every groundbreaking discovery there are hundreds of examples of people poisoning or severely injuring themselves. Take for example Paul Karason, who started regularly ingesting a compound called colloidal silver in the hope of relieving certain ailments, and turned his entire body permanently blue, earning himself the nickname 'Papa Smurf'. Be extremely careful about what you put on or into your body, read warnings on packaging, follow medical advice and, if at all in doubt, don't do it.

CHAPTER 1

Let Us End at the Beginning: Ageing and Death

'We all know that our time in this world is limited, and that eventually all of us will end up underneath some sheet, never to wake up.'

Lemony Snicket, *A Series of Unfortunate Events: The Reptile Room*

Ageing inevitably ends in death. But what is death? Defining death is surprisingly tricky. One definition states, 'Death occurs when there is permanent loss of capacity for consciousness and loss of all brainstem functions.'[1] All roads eventually lead to brain death, sometimes via stopping the heart. Basically, if your brain stops working you die. This can happen either because the brain itself stops working (Figure 3) or because the blood flow to the brain stops. But just putting 'brain stopped' as a cause of death doesn't help statisticians use relative causes of death as a way to plan and prioritise prevention. This explains the need for seventeen thousand ICD codes: looking at the proximal cause of death can be informative.

One way to visualise causes of death is to see the organs that failed on display. In general we keep our organs on the inside – an eminently sensible approach – but we can

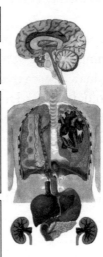

Brain: <u>Stops working – you die</u> (All death)	**Dementia:** Stops lungs or heart <u>No blood to brain – you die</u>
Stroke: <u>No blood to brain – you die</u>	
Cancer: Tumour spreads to other organs, causing them to fail. Ultimately leads to either <u>no blood to brain or brain not working – you die</u>	**Heart:** Stops pumping <u>No blood to brain – you die</u>
	Lungs: No oxygen <u>Brain stops – you die</u>
Liver/ kidney failure: Messes up blood. <u>No blood to brain – you die</u>	**Diabetes:** Causes heart to stop. <u>No blood to brain – you die</u>

Figure 3: **All roads lead to brain death. The actual and proximal causes of death.** How failure of different organs leads to brain death.

occasionally glimpse their inner workings. Like most medical schools, Imperial (where I work) hosts a Pathology Museum cataloguing the multitude of ways in which our organs can fail us (or that we fail them). It really rammed home the terrible things that can happen to your body. On entering the museum, a whole baby's arm pickled in formalin beckons you in, followed by shattered bones, soot-black lungs, enormous tapeworms and so, so many brains in jars. The collection contains an extraordinary, nightmarish, (no longer) living history; its notable *objets de mort* include a bone shattered by a musket ball at the Battle of Waterloo, seven bedsprings from the stomach of a Rhodesian jail inmate, the obligatory rubber truncheon recovered from someone's bottom, a uterus from the first hysterotomy performed in the UK and a stomach-shaped hairball from a girl who died because she ate too much of her own hair.

The collection was curated not by a retirement-aged Igor, replete with lightning and cobwebs, but by a tall, fresh-faced archaeology graduate, Margaret Lever. She pointed out that jobs in archaeology are hard to find, with opportunities for freelance tomb raiding restricted to the movies. Whilst bones remained her main passion, Lever was learning the trade of preserving organs thanks to *Medical Museum Technology* by J. J. Edwards and M. J. Edwards, published in 1910. An invaluable guide for curators and serial killers alike, it contains handy notes on liquid mounting, colour injection, the use of plastics, maceration and articulation of bones. When I asked if it felt a bit odd having your lunch next to a room full of body parts, Lever retorted, 'If you think this is odd, you should see the dissection floor upstairs!' I passed. The museum was collated by organs, like so much fleshy Lego, building blocks of people long gone, scarred and blackened by their lives. And Imperial holds just a small collection – the largest collection belongs to our neighbours (and they think rivals) King's College London. However, due to complexities of the Human Tissue Act, neither collection is open to the public. If you live in the UK and really need to see a brain in a jar, head to the Hunterian Museum, London, whose collection is sufficiently old that they can legally share it with the public.

Alongside death,* laws are the other things that we cannot avoid. And you might imagine that a legal definition of death exists. But the UK lacks a statutory definition of death; instead, the courts widely accept the neurological

---

* And taxes, obvs.

criterion – death of the brainstem. As a side note, in the UK, when you die, no one *owns* your body either. Death is quite murky in UK law and when we pass is predominantly a medical rather than a legal decision. This affects end of life and cessation of support. British law distinguishes between actions that deliberately end a life (e.g. overdose of morphine) and the removal of life support – even if that inevitably leads to death. Ending a life deliberately constitutes murder; removing life support does not. Medical decisions to withdraw life support are based on whether the life support benefits the patient more than it harms them. Originally, only judges could decide whether support could be removed, but since 2018 the decision can be made outside the courts if the family and the doctors are in agreement. Judges still intervene in split decisions.

But I plan to avoid the tender scene of my nearest and dearest arguing over who gets to pull the plug on me for as long as possible. Barring some acute trauma, my most likely route to the end of life is that one or other organ gives up on me (ultimately leading to brain death – see Fig. 3). Whilst death marks the end of the road, the final furlong carries equal weight. And it's here that ageing plays a significant part; I have been desperately hoping to ward off ageing's spectral presence ever since I found that first true grey hair.

CELLULAR AGEING

If defining death is tricky, defining ageing is far harder. Ageing of the cell plays a central role in our becoming older. As you hopefully recall from school science, our bodies are made of cells. Cells are the basic unit of life. Ignoring viruses, anything biological that is smaller than a cell is not alive.

The cellular conception of life dates back to the 1660s, when the cell was first identified by Robert Hooke, one of those irritating/extraordinary seventeenth-century polymaths who discovered a huge range of fundamental things. You might remember him from Hooke's Law, loading an elastic band in physics until it goes twang. When not busy inventing physics, Hooke also developed one of the first microscopes, publishing his observations of plant cells in 1665. Hooke's studies eventually led to cell theory, developed by Robert Remak, the tenets of which are: cells make up all organisms and all cells come from other pre-existing cells. The final point was neatly coined in Latin as *Omnis cellula e cellula* by Rudolf Virchow (who will feature again in our narrative; for now just remember sausages, it'll make sense later).

Some organisms, like bacteria, comprise only one single cell. These single-celled organisms make up the vast majority of biological diversity on Planet Earth, living as they do in environments such as hot springs, frozen lakes and everything in between – some even thrive on the walls of nuclear reactors. The more complex multicellular organisms, like us, encompass a very small sliver of the breadth of possible life. However, multicellular organisms account for most of the living material on the planet (called biomass), or at least plants do. Plants account for eighty-two percent of all biomass on earth; humans contribute a tiny 0.01%.[2] Still, we far outweigh the nematodes' 0.004%; take that, nematodes. But whilst nematodes contribute a puny 0.02 gigatonnes of carbon to the whole earth ecosystem, they have contributed a significant amount to our knowledge of ageing, because they live a conveniently short time.

Complex organisms, like us, are comprised of many different cell types, all performing different functions;

15

several cells doing the same thing come together as a tissue, and several tissues working together for a grander purpose form an organ. In spite of their functional differences, all cells share a core cellular machinery that they require to operate. They need to be able to take their genetic information (stored on DNA), turn it into the messenger molecule (RNA) and produce proteins. For this they need the ability to convert glucose into energy via the mitochondria, the powerhouse of the cell, which can perform the requisite chemical reactions. Cells also sense their surroundings, talking to other cells both near and far. But from the day we are born, our cells slowly lose functionality. As cells age, they gather hallmarks of senescence, losing the ability to perform underpinning core tasks. You can afford to lose a cell here or there, but eventually the cumulative damage tells, and ageing cells lead to ageing organs.

As I saw in the Pathology Museum, humans have found an impressive array of ways to damage their organs. When younger, we can recover some damaged organ function through replacement with other cells. Replication and repair are restricted to a specialised subset of cells, called stem cells. Whilst nearly all cells contain a complete copy of our DNA, not all cells can read its entirety. It is like getting into a library but discovering restricted sections hide behind *Doors of Stone.** The function of the cell is determined by which parts of the DNA library it can enter. This lets them specialise: a cell in your retina that receives the light to read these words is substantially different from one in your brain that turns them into my voice inside your head. Stem cells possess an access-all-areas pass, letting them become

---

* That we may never get to read.

anything they want; this enables them to replace damaged cells. As they age our stem cells lose the ability to replicate, and damaged regions can't be replaced. Stem cell exhaustion is one of the nine hallmarks* of ageing.[3] Failure to replace damaged tissue contributes to most common causes of death.

Why do our cells age? To find out, I talked to Dr Cathy Slack, associate professor at the University of Warwick whose expertise pertains to all matters ageing. Cathy and I did our PhDs together at Imperial, me on tobacco plants, she on fruit flies. As students in the early 2000s, we spent a lot of time in her lab watching Robbie Williams videos on the internet. Some things stick – I never left Imperial, Cathy still works on fruit flies; some things don't – neither of us likes Robbie Williams anymore. I caught up with her as she recovered from 'the lurgy' as she put it (she's not an infectious disease expert like me, otherwise she might have said something more technical, like a *stinking cold*).

Cathy told me about two main theories of cellular ageing – programmed and damage-related. They are often seen as competing (mainly because that is a good way to increase your own research income at the cost of the other camp), but almost certainly both contribute. The programmed cell ageing camp has its roots in the work of Leonard Hayflick, who demonstrated in 1968 that, outside the body, human cells have a finite lifespan. Until his work, the prevailing theory posited that you could continue growing cells in a

---

* The others are genomic instability, telomere attrition, epigenetic alterations, loss of proteostasis, deregulated nutrient-sensing, mitochondrial (powerhouse) dysfunction, cellular senescence, and altered intercellular communication.

glass jar indefinitely – this is called *in vitro* after the Latin for 'in glass'. Hayflick showed that, after about thirty rounds of division, cells give up the ghost. This limit of replication is not universal. Somewhere in Eastern California towers Methuselah: a 4,855-year-old bristlecone pine tree, the oldest living thing. Ming the Clam holds the record for the oldest living animal; found off the shores of Iceland, it reached 507 years. But I question its quality of life – yes, five hundred is a lot of years, but if your day only involves slurping plankton from the North Atlantic, it hasn't got that much going for it. Though no doubt the biohacker crowd would be jealous.

The widely received cap on human lifespan is 125 years; if I make this milestone, I would have seen three centuries. The oldest recorded human (with a verified date of birth), Jeanne Calment, lived to 122, famously smoking till she turned 100, drinking port and eating a substantial amount of chocolate. The more optimistic gerontologists think that human lifespan lacks limits, and that among us there might already be people who will live even longer than Calment. Slack firmly believes human lifespan has a cap, telling me cells contain innumerable biological constraints: 'It's like the hydra – you fix one ageing process only for three others to crop up.' Most scientists I spoke to thought the same, arguing that increasing healthspan – the amount of time we are healthy – rather than just increasing lifespan should be the goal.

One marker of programmed cell ageing is the telomere, a component of our DNA. Telomeres bracket our chromosomes, capping the long strands of DNA like the protective plastic at either end of a shoelace. Bookending the chromosome makes the DNA more stable, facilitating accurate replication for when cells divide. With each round of replication, telomeres shorten and deplete, until below a key threshold, the chromosome's readability and replicability is compromised, and cell replication stops. The rate of telomere loss varies across different species and may correlate with lifespan. Dogs lose DNA faster than humans, and breeds of dog with shorter telomeres live shorter lives; the long-telomere Beagle outlives the short-telomere Bulldog.[4] The ratio of DNA loss may even explain 'dog years' – canines lose DNA nine times faster than humans and live nine times shorter lives. Telomere lengths reset into the next generation; fortunately for them, we don't pass on our cellular age to our children. But just because our telomeres shorten over time doesn't mean they are the sole drivers of senescence. 'There has been a bit of a failure of communication around ageing,' Slack says. 'Telomeres are just one part of the process'. This reflects the morass of information I was to discover around living longer; the complex process of ageing gets simplified down to a single factor or intervention – extend your telomeres, extend your life. Back to Slack: 'telomeres are writ large in the public imagination as being central to ageing, way more than their actual impact'. Extending your DNA and therefore lifespan with telomerase makes for a nice journalistic narrative, but the experimental data does not support it.

BORN TO DIE

Why have we evolved to age? What possible advantage could there be? As Slack says: 'From an evolutionary perspective, ageing doesn't make sense. You shouldn't evolve a process that's bad for you and ultimately leads to your death.'

Evolutionary biologists have struggled for the last seventy years to square this particular circle; one theory, which even Slack admits is 'a bit of a mouthful', is antagonistic pleiotropy. It means that genes selected for fertility and growth may also contribute to ageing and death. Evolution selects at the point of reproduction, not for longevity – if you beget more children but die young you will pass more genetic information to the next generation than if you procreate less but live longer. This particularly had an impact in the mists of ecological time when other factors stopped you living longer – such as sabre tooth tigers, smallpox and starvation. If you're too old to reproduce, evolution (mostly) doesn't care about you. Slack compares antagonistic pleiotropy to filling a bath: 'You need to turn the tap on to get the bath full, but if you leave it on too long, the bathroom floods.' Ageing isn't the only thing where the net result of evolution doesn't make sense; much of the male human body is poorly ergonomic: non-functional nipples, testicles on the outside,* the passage of the urethra through the prostate, the appendix.[5] Evolution is after all a random process of trial and error, not design.

Damage to the cell is then layered on top of the programmed cellular ageing. Slack compares the damage model to an 'old banger of a car' – eventually the rust takes over and the car (our body) falls apart. As a long and extremely complex

---

* A surprising number of competing theories (unsurprisingly postulated by men) try to explain the source of the scrota; suggested reasons include: cool-storage, sperm training, a display of sexual health and, hilariously, to enable galloping (try running in the nude to disprove this one). It should be noted that not all mammals wear their testicles on the outside – you'll never see an elephant's testicles, you just have to imagine the size of his balls!

molecule, DNA is particularly susceptible to damage, and because it gets passed down from cell to cell, genetic damage accumulates over time. This can occur in the form of point mutations, little changes in the sequence – imagine a photocopy of a photocopy a thousand times over. One type of alteration affecting gene function relates to the way in which cells store DNA. Thanks to Franklin, Crick and Watson, we carry an image in our heads of DNA as beautiful, intertwined spirals, but it forms far more complex structures. DNA winds itself around proteins called chromatin, like string spooled around beads. This avoids it getting jumbled up like Christmas tree lights thrown carelessly into a box till next year. However, random winding of DNA around the chromatin bobbins would make it difficult to find the right gene in a timely manner. To resolve this problem, modifications in the DNA backbone called epigenetic markers switch genes on and off. These markers respond to environmental cues; in essence our cells become trained to react faster. However, epigenetic modifications on our DNA increase with age, impacting the way the cells behave. Fortunately, as with telomeres, epigenetic markers reset at birth. But it's not just DNA that is prone to damage: cells accumulate other toxic biochemicals, particularly misfolded proteins; all of this toxic guff leads to cell death, which can then in turn trigger a cascade of local damage via the immune system.

IMMUNOSENESCENCE: MISFIRING OF THE BODY'S DEFENCE SYSTEMS

Like the rest of your body, your immune system ages; evolutionarily it is more important to protect younger members of the species from infection, enabling them to have more babies, than to protect those past reproductive age who will not spread their genes any further. As we age, two things happen to our immune system, both bad. We lose the ability to fight off infections, and low-level inflammation increases. A similar issue causes both problems – the loss of immunological specificity. The immune system needs extremely tight regulation; an out-of-control immune system is as dangerous as no immune system at all. This is why it pays to be extremely cautious about 'immune-boosting' supplements. You do not want your immune system boosted non-specifically. An infamous clinical trial used a drug (TGN1412) that switched on the whole immune system, with disastrous results for the volunteers.[6] What we need as we age is better immune regulation; easy to say, but harder to envision what it actually would involve and how you can affect it through drugs, foods or lifestyle changes.

To understand more about immunity and ageing I spoke to Professor Luke O'Neill of Trinity College Dublin. O'Neill is the definition of a polymath, being not only an author, radio personality and scientist but also lead guitarist of a covers band – The Metabollix. I caught up with him at an immunology conference overlooking the Acropolis in between his science talk and him laying down his best Guns N' Roses riff.

O'Neill told me that, in terms of ageing, the immune system's appetite for destruction is driven by inflammation. Early in your life, inflammation is a good thing. It signals to the immune system that there is a threat requiring a response. If a splinter pierces your skin, the surrounding tissue goes red (*rubor*), heats up (*calor*) and begins to swell (*tumor*); these being three of the four classical symptoms of inflammation, the fourth being pain. The redness and swelling reflect the opening of blood vessels to allow in immune cells; the heat is thought to kill bacteria. Whilst, as the name suggests, pain is painful, we need it to initiate repair; nerve cells that detect pain also recruit immune cells that fix damage. As an acute response, inflammation is beneficial. Problems arise when the inflammation doesn't resolve. All of the big hitters – heart disease, stroke, dementia, cancer, lung disease have inflammation as a causative element. Inflammation accelerates negative processes, increasing damage to DNA and cells and reducing our ability to replace them. Unresolved inflammation can lead to the accumulation of cells in the wrong places – in lung disease that can reduce our ability to breathe, in heart disease they can block the crucial blood flow to the heart. O'Neill explained how inflammation drives a range of brain diseases including Alzheimer's and Parkinson's. In these diseases, proteins build-up in the brain, 'irritating and frustrating' immune cells, causing them to explode in a process called pyroptosis, leading to further damage.

Signalling molecules in the blood give us a window into our body's inflammation. The immune system talks to itself and other cells through proteins called cytokines. They are, or should be, strictly controlled to prevent off-target damage. Daniela Novick discovered some of the first cytokines in an unusual source material – the urine of post-menopausal nuns. In a talk I attended, she described this as a goldmine – one nun's trash is another scientist's treasure. Remarkably the company Serono was already collecting holy pee in the Holy See on a mass scale in order to make the fertility hormone Pergonal. It took ten nuns ten days to make enough urine for one treatment; nuns' urine being particularly valuable because pregnancy can inhibit fertility.

The first baby fuelled by mother superior's micturition was born in 1962; by the 1980s, the company needed thirty thousand litres a day to keep up with production,* eventually switching to other methods as the supply dried up.

The cytokines Novick discovered in nun number one play a central role in ageing-associated damage. They can stimulate cells into a destructive pathway; for example, in osteoarthritis, they encourage cells to break down the collagen that connects tissue together. Immunologists have tried to develop a grand unified theory that links ageing, damaging and the immune system, coining the inspired portmanteau 'inflammageing'. This theory owes as much of its popularity to its snappy name as to what it shows, which is a bit hand-wavy.

NEGATIVE INPUTS

Gout, with its characteristic joint tenderness, particularly in the toes, exemplifies how misfiring inflammation causes disease. In gout, a substance called uric acid (a breakdown product of DNA) accumulates in the joints. Normal, healthy tissue does not contain uric acid, and the body treats its presence as a danger signal. The accumulation of uric acid in the toes confuses the immune system into thinking itself under attack, initiating a cascade of inflammation in the pinkies, causing pain. Gout, described as the 'disease of kings and the king of diseases', has been around since antiquity. Whilst diet isn't the only cause, foods high in purines, such as anchovies or Marmite (ensuring vegans

* The average output is three hundred millilitres, so this amount needed 100,000 trips to the bathroom. That's a lot of nuns.

don't miss out) can trigger gout, with alcohol consumption also a contributing factor. Incidentally, whilst Marmite might seem the most British of spreads,* it was invented by a nineteenth-century German polymath, Justus von Liebig, who also invented the Oxo cube, fertilisers and the condenser (which you might have used during GCSE chemistry to separate liquids by distillation); he also sowed the field of organic chemistry.

Inflammation accelerates organ ageing; through it, the most well-known risk factors (smoking, overeating, boozing) make us ill. Obesity causes general systemic inflammation, 'putting the cells into overdrive', as O'Neill put it. Exposure to air pollution and cigarette smoke drives lung cell death, leading to inflammation. The elevated sugar of diabetes has the same impact on immune cells as a bag of Haribo has on a small child; it supercharges them, ending in tears. Alcohol kills the liver, releasing toxins leading to (guess what) inflammation. Stress (via cortisol) stimulates systemic inflammation. Our lifetime burden of infections also plays a role. Chronic viral infections keep the immune system on edge, never allowing it to relax, adding to the inflammatory soup of stimulation. And it isn't just viral pathogens: the 'good' bacteria that we carry on and in us, the microbiome, can contribute to inflammation. An excess of bacteria in the mouth and the gum disease they cause is linked to heart attacks. Buccal bacteria can enter the blood stream via rotting teeth, turning up the thermostat of inflammation. As we will

---

* Vegemite (the Australian Marmite) was developed in the 1920s in response to a shipping crisis caused by the First World War which prevented Australians from getting their delicious, yeasty, toast-delivered goodness.

keep seeing, sustained inflammation during ageing is problematic especially when linked to stress.

SCIENCE THE HECK OUT OF THIS

How have medical researchers linked the diseases of ageing to inflammation? The scientific process linking cause and effect has three building blocks: scientific mechanisms, disease models and epidemiological studies. Scientific mechanisms are proposed systems of how things work; in the medical world, they are theories of how disease impacts the body, which can be tested in clinical settings. Disease models serve as the sandbox in which the mechanisms can be tested. They often use small (and not so small) animals, enabling scientists to evaluate aspects of the mechanism in isolation, building up a complete picture. Disease models are by necessity limited – the life experience of a lab mouse is not the same as an ageing human. The statistician George Box coined the great adage 'all models lie, some are useful'. Epidemiological studies are the final step; they allow researchers to analyse disease throughout large human populations. Indeed, the word 'epidemiology' derives from three Greek words – *epi* (on), *demos* (people), *logos* (study). An ideal study compares people matched as closely as possible except for the variable we want to test.

To put this into context, our scientific mechanism might propose that inflammation causes heart attacks by damaging the blood vessels feeding the heart. Then, in our disease model, we would induce inflammation in the blood vessels of mice and determine whether this causes heart damage. Finally, we would conduct an epidemiological study, looking for people with and without blood vessel inflammation and measuring the frequency of heart attacks among each group.

THERAPY?

Identifying the cause of a disease opens up the path to developing treatments. But not all proposed treatments elicit the effects their proponents claim. The development of new (evidence-based) interventions requires a similar thought process to identifying the cause of a disease. Taking a critical thinking approach is *really* important in how you approach any suggested wonder drug/intervention/crazy idea that you see on the internet suggesting it will make you live forever. One approach is to consider the three aspects that an intervention needs to be robust: a rational hypothesis, a plausible mechanism of action and then proof of effect in clinical trials. I recommend that you apply this framework to everything before you try it, as I won't always be on hand to tell you, 'No, eating nothing but hand-fermented yak yoghurt will not do wonders for your diabetes, but yes, going for a walk will.'

STEP 1. RATIONAL HYPOTHESIS

The word 'hypothesis', from ancient Greek, means 'to put under scrutiny'. There are many working definitions of 'hypothesis', over which philosophers spend much time and effort arguing, sometimes with pokers. You can spend a career in actual (rather than theoretical) science without ever stating an opinion about the nature of hypotheses. If forced, I would say that I draw on the ideas of empiricism from Francis Bacon and use an adaptation of Karl Popper's:

a hypothesis must be falsifiable.* Falsifiability plays a critical role; to be science, it should be possible to disprove using a counterhypothesis with better evidence.

In the case of an intervention that reduces ageing by reducing inflammation, we need a testable hypothesis with a measurable outcome. Let's take a real-world example using the first thing that came up when I googled anti-ageing foods: tomatoes. The question is therefore: 'Do tomatoes make you live longer?' This can be rephrased as a hypothesis: 'Tomatoes reduce ageing diseases because they reduce inflammation.'

STEP 2. PLAUSIBLE MECHANISM OF ACTION

Based on current knowledge, can the intervention do the thing proposed? Tomatoes can be linked to reduced inflammation through the compounds that make them red. Specifically, lycopene is an antioxidant and can inhibit a key immune-signalling pathway that leads to inflammation (NF-κB). Therefore, there is a somewhat plausible mechanism – though admittedly a long way from proving that ketchup keeps you young.

STEP 3. CLINICAL TRIAL

To demonstrate effect, there needs to be one or preferably several clinical trials to prove the hypothesis. Humans have

---

* I also like to throw in a bit of 'Occam's razor' – that the simplest answer is always the best. For example, Lee Harvey Oswald killed JFK, because when you put all the evidence together, it is the simplest and therefore the best answer.

been performing trials on food for as long as there have been humans – at some point in the distance past the first brave (foolish) soul would have prised a shrivelled black mushroom from the jaws of a pig and decided it would taste great on pasta. The first true clinical trial is widely attributed to James Lind, surgeon aboard the HMS *Salisbury* in 1747. During the War of the Austrian Succession, the Royal Navy were up to their usual business of blockading the English Channel from those pesky continental Johnnies. After eight weeks at sea, twelve of HMS *Salisbury*'s sailors developed scurvy (due, as we now know, to a lack of vitamin C). Lind split the sick sailors into six groups of two and gave them (somewhat vile-sounding) different treatments – as follows:

1. 1 l cider;
2. 25 ml of elixir vitriol (dilute sulphuric acid);
3. 18 ml of vinegar three times daily;
4. 280 ml sea water;
5. Two oranges and one lemon;
6. A medicinal paste made of garlic, mustard seed, dried radish root and gum myrrh.

By day six, the fruit group (group 5) had returned to work, singing shanties and engaging in the sailors' standard routine of rum, sodomy and the lash; whilst the rest continued moaning, vomiting and losing teeth. Critically, Lind took people with the same condition and gave them different interventions, allowing him to compare the impact.

Clinical trials have increased in sophistication since Lind's lemons. One important innovation was randomisation, assigning participants to groups by a computer, not the study physician who could inadvertently bias outcomes by choosing healthier people for the active arm (the participants receiving the drug or treatment). Another innovation is blinding, where participants don't know what they are receiving, since knowledge of treatment can greatly affect results. In Lind's study, the group eating the citrus fruits would have known they were not drinking half a pint of sea water, but this didn't matter because the outcome (curing scurvy) occurred independently of the sailor's behaviour. The challenges of blinding are particularly great in behavioural studies – it being extremely difficult to hide from someone whether or not they belong to the singing group in a trial investigating the benefits of choir membership on lung function.

For some outcomes (particularly subjective ones, like pain), knowing you are in the study's intervention group can alter responses. For example, if you tell study participants that the red pill will make them happier, a proportion of them become happier regardless of what it contained. This is the famous placebo effect. The placebo takes the form of a control treatment that looks like the actual drug but lacks active ingredients. The act of taking a pill alone can be enough to improve health, even if the pill contains nothing more than sugar. Remarkably, in one study, even when volunteers knew that they were taking inert pills, they reported improvements in irritable bowel syndrome, the placebo effect being so strong.[7]

The placebo effect is an example of the complex interplay between mind and body in clinical trials. People volunteer for clinical studies for complex reasons; strong belief about an intervention can act as a motivating factor. For example, if you really believe in the power of blueberries to avert ageing, you might be more likely to join (and stay in) a study exploring their effects. The power of your belief might then affect the outcome of the study. The converse can also happen: if a study tests something everyone 'knows to be true', recruiting volunteers becomes difficult. One study explored this by offering volunteers the chance to jump out of a plane with a parachute or a placebo control backpack; surprisingly, twenty-three out of the ninety-two people screened volunteered.[8] Whilst the actual study was a bit silly (the plane was on the ground),* the thinking behind it was important – why would you volunteer for a trial where the intervention so palpably works? The placebo effect shares a darker sibling, the nocebo effect, where expectations of side effects make them worse.

In the case of our tomato experiment, we could take two groups of people and measure the effects of tomatoes on age-related diseases. Easily said, but hard to do well, given all the other little differences between people that mask the effect of the tomatoes. The ideal study would take as many identical twins as possible (which removes the complication of genetics) and feed one from each pair an extra tomato a day. For the perfect experiment, each pair of twins would also need to live together, work the same jobs and, apart from the bonus tomato, eat the same food at the same time on the same days. Having set up the study, we would simply follow them until they died. If the tomato group lived longer (on average) we could conclude that tomatoes were indeed the causative factor. This is obviously impossible. These challenges explain why scientists perform animal studies

---

* The authors drew the excellent conclusion that 'the trial was only able to enroll participants on small stationary aircraft on the ground, suggesting cautious extrapolation to high altitude jumps.'

– you can control for the other variables. But, to quote my other favourite science adage, 'The best model of a cat is a cat, preferably the same cat.'[9] Animal models have limits when translating to humans.*

When investigating ageing, time and money constraints mean most studies will not follow an intervention all the way to death. Compromises are needed. The more compromises made, the less robust the conclusion. Sometimes, instead of measuring the final outcome, we measure a correlate – something found alongside your condition but that doesn't necessarily cause it. To demonstrate a protective effect of tomatoes, we could measure proteins linked to inflammation in the blood. One study did just that – it compared levels of a marker of inflammation called TNF in two groups of patients with metabolic syndrome: one group drank tomato juice, one didn't.[10] The authors observed significantly less

---

* A note on the value of animal models. Humans are unique and individual as snowflakes. The degree of complexity between people is so great as to make many comparisons meaningless. With animals, we can control many more of the variables. For example, the commonly used white lab mice are genetically identical and housed in the same environment, eating the same food. This gives us confidence that the intervention we perform is having the effect we attribute to it (unless, of course, they are actually hyperintelligent pandimensional beings experimenting on us). Animal models form a critical part of the science ecosystem for the development of ideas and for demonstrating that new drugs will not instantly kill people. A lot of work goes into reducing, refining and replacing animal models to ensure the harm is minimal and justified. All animal work in the UK is performed in a tight legal framework, curiously falling under the remit of the Home Office. Interpreting the results carefully and not over-extrapolating is vital: as my old boss Professor Peter Openshaw once said, 'Mice don't lie but sometimes we misinterpret their squeaks!'

TNF in the tomato juice-swilling group. However, before you race to fill your basket with Bloody Marys, take careful note that the study only enrolled twenty-seven volunteers.

The number of participants is a key determinant for interpreting the value of any research study. The more people included in a study the better, because people are so wonderfully varied. Averaging across lots of people improves predictive value. The predictive ability of a study based on its size is called the power. A study with too few people is described as underpowered. To increase the statistical power, studies can be combined with other similar ones: a meta-analysis. For tomatoes, a group at Northumbria University scanned 3,970 scientific papers that mentioned the word tomato and heart disease, but rejected 3,945 (presumably because they were so unspeakably bad no one should ever read them). When combined, the data from the twenty-five remaining studies encompassed 211,704 participants[11] and showed that participants with higher lycopene intake had significant reductions in stroke risk and mortality. Eat more tomatoes, problem solved.

If only. To paraphrase the American journalist H. L. Mencken: 'For every complex problem, there is a simple solution, and it is wrong.'[12] Every cell in our body is extraordinarily complex, so complex that my mind boggles about anything happening in it at all. These immeasurably complex cells then make tissues, that make organs, that make bodies which possess a mind of their own, interacting with the world in a vast and unique way. Simply put, ageing is complex. Which is almost as equally as unhelpful as saying the secret to healthy ageing is to eat more tomatoes.

There needs to be a middle way, a path that reflects the complexity of the system but can be translated into things to improve our lives. One approach to avoid diseases of ageing is to avoid the factors that cause them in the first place. Fortunately, health, ageing and disease are all inter-connected; altering behaviour to reduce one disease, e.g. heart attacks, will also reduce risk of strokes and dementia. But the flip side also applies, conditions stack; if you have a heart attack, a slew of other conditions most likely follow close behind.

And yet, we can all recall examples of people who smoke and drink their way into their hundredth year of life. Environmental exposure is extremely important, but it doesn't account for everything. Longevity tends to run in families, suggesting an inherited element, potentially explained by genetics. This gives me the first opportunity to self-experiment and hopefully unlock the secret to better ageing – read my genes.

CHAPTER 2

The Fault in our Genes

'DNA neither cares nor knows. DNA just is. And we dance to its music.'

Richard Dawkins, *River Out of Eden:*
A Darwinian View of Life

Dissecting the mechanisms of ageing and death did not make me any more positive about the inevitability of my decline and eventual demise, so instead I turned to science to see what the future might hold. Firstly, I needed to identify which of my organs was most likely to mess me up in the future. Since, to paraphrase Larkin, my parents filled me with their faults, where better to start than with inheritance.

Once upon a time, the only way to anticipate your fate was to look back to your predecessors. This is imprecise, at best. Mainly because of the huge changes in risk and exposure in our respective lifetimes. The landscape of pathogens that assails us now is so completely different from that of one hundred years ago as to make comparisons almost meaningless. Even if your grandparents avoided killer pathogens, their lifetime toxin ingestion would have been extremely different from yours. You are far less likely to smoke, work in a mine, use toxic chemicals at work, breathe in smog or drink polluted water than people born one hundred years before you. Likewise, you are much more likely to eat fresh vegetables, retain most of your own teeth and have access to antibiotics. In my immediate family of four, three of us would be dead without modern medicine: me (cancer), my son (infection) and my daughter (meconium inhalation at birth).

All this means that knowing that my grandparents died of a stroke (a legacy of the Second World War), pneumonia (because of smoking) and vascular dementia (linked to smoking) doesn't really tell me much because I avoided fighting a brutal winter campaign in the Italian mountains and don't smoke. The astute amongst you will notice that only accounts for three out of four grandparents; the other one, my paternal grandmother, died of 'old age'. Which is equally uninformative. The good news is that, anecdotally, longevity runs in my family. My great aunt lived to the ripe age of ninety, but I suspect a degree of familial cherry picking – most families can point to a distant relative somewhere who broke the one hundred mark and another who died young.

We can now use an alternative approach – gene sequencing, which has progressed extraordinarily in the last twenty-five years from the first 'completed' human genome.* Since that landmark event, sequencing has changed from a multi-million-dollar, multi-partner global consortium to a relatively low-priced commercial kit for use at home. The cost of

---

* Whilst the commonly reported date for completion of the first human genome is 2003, remarkably, the complete read of the human genome was only finished in 2022. The 2003 version contained ninety-two percent of the sequence, including all the actual genes; however, a lot of our DNA is highly repetitive. The techniques used to read DNA struggle with repetition, so it took another twenty years to finish off characterising the three billion sequences.

genome sequencing dramatically dropped from $100 million in 2001 to $1,000 in 2023; i.e. from that of a spy satellite to an iPhone. The logarithmic price drop makes gene sequencing far more routine, both in labs and hospitals.

What does all of this genetic information mean and how does it affect our health?

MONKS AND PEAS

Genes are the fundamental unit of inheritance. They serve as the instructions by which cells are made. One gene encodes one protein. Proteins in their turn make up our cells. And whilst our cells comprise more than just protein, the proteins do the heavy lifting – they perform structural, signalling and synthetic functions. Proteins build the cells, the cells the tissues, the tissues the organs and the organs the person. Therefore, the instructions for proteins serve as the instructions for us. What boggles the mind is that everything and everyone that lives and has ever lived uses the same genetic code to make their bodies. Look around you: all the living things you can see use the same underpinning software. And yet in spite of using the same code and the same building blocks, there is enough variety to encode both sunflowers and sun bears. Even within species, sufficient combinations exist to ensure that (identical twins asides) no two people are exactly alike. It's one of those things that is so astonishing that it is best not to think about too deeply.[*]

Gregor Mendel first developed ideas of discrete units of inheritance with his famous pea plants. Born Johann to a

---

[*] Alongside how the universe can both be infinite and expanding, or the inexplicable attraction of modern jazz.

poor Czech farming family in 1822, he became a monk to continue his education, upgrading to the name Gregor in the process. Being a monk probably helped his scientific progress, as the experiments are by necessity slow – following multiple generations of the humble pea. I certainly lack that level of patience – pestering my lab team to share their results with me as soon as their experiments are finished, if not sooner. He studied seven different traits in the pea: seed shape, flower colour, seed colour, pod shape, pod colour, flower position and plant height. The traits behave similarly enough between generations that we can use flower colour as an example to explain the rules of Mendelian inheritance. Mendel compared two colours of pea flower, red and white. Before Mendel, scientists believed that offspring represented an even mix of their two parents, and therefore that mixing red and white pea plants would give you pink – called blended inheritance. Mendel disproved this. He cross-pollinated a red pea with a white one, and the first-generation (sometimes called F1) offspring *all* flowered red. However, when he then interbred his F1 pea plants (pea incest not being a thing) the offspring *did not* all flower red: for every three red plants there was one white plant. This indicated the existence of two copies of the gene determining flower colour and that they segregate independently. The genetic make-up of the plant (or us) is called the genotype, the invisible engine that determines its characteristics. The external visible characteristics are called the phenotype. The interactions of the genes determine the phenotype. Mendel described the gene that won out in the phenotype – in this case red – as dominant and the masked gene recessive. This terminology is still used for binary traits. A question lingers about whether Mendel fudged his results;

this isn't helped by Mendel insisting on his data being burnt on his death.* According to the geneticist and mathematician Ronald Fisher, the experimental data is a bit too perfect, the results should be a bit noisier; but statisticians in skewed distributions shouldn't cast outliers. Fisher's own legacy is more than a bit eugenics-y.

D, R AND OTHER NAS

Whilst Mendel discovered the underpinning theory of genes moving from one generation to the next, the molecular mechanism remained unclear. It was appreciated that the cell nucleus housed genetic information, but it took a long time to demonstrate that DNA was the molecule of inheritance. DNA's role as the key molecule came through the work of many people, none of whom was named James or Francis or even Rosalind. Oswald Avery anticipated their work by showing that nucleic acid was the stuff of which genes are made by adding single components of bacteria back into a culture one at a time, demonstrating that traits only transmitted after the addition of DNA. Alas, the pathway to DNA fame is littered with people who didn't get the recognition they maybe deserved. Avery received Nobel nominations almost yearly from the 1930s but won nothing, while Rosalind Franklin missed out through a combination of sexism and death.†

---

* Hiding your workings was very much a nineteenth-century scientist thing to do: Pasteur told his family to never reveal his notebooks and his papers remained restricted until 1971, eighty years after his death.
† Being alive is a criterion for getting a Nobel, with one exception: Ralph Steinmann, who sadly died between the decision being made and announced.

DNA comprises four different base pairs – usually represented as the letters A, T, C and G: standing for adenine, thymine, cytosine and guanine. In the double helix, DNA strands complement each other; A always pairs with T and G with C. When a cell divides and the two strands separate, they can be rebuilt with their exact matches. For the DNA's instructions to be used to make proteins, a smaller molecule called RNA makes a copy of part of the genetic code and brings it to the ribosomes, the cell's factories. Each triplet of DNA letters (called a codon) specifies one amino acid, and the order of codons informs the order of the amino acids. Put amino acids together in particular combinations and you get whole proteins. There are twenty different common amino acids, and loosely they can be grouped by whether they have a positive or negative charge and whether or not they repel water. These characteristics stack and the amino acids that make up a protein determine its form and therefore its function (Figure 4).

There is a lot of molecular science buried in the statement 'when the two DNA strands separate, they can be rebuilt', involving many proteins ending in the suffix -ase. The complexity of the process that enables DNA replication contributes to another feature of genetics – mutation, the driver of evolution. Any change in DNA sequence can be considered a mutation. Small changes in the order of the amino acids can make huge changes in the function of the protein they make. For example, as well as encoding amino acids, three sets of DNA triplets encode STOP codons; these do exactly what they say in the code – tell the ribosome to stop making the protein. Mutations that change an amino acid codon to a STOP codon will have pronounced effects, causing the untimely termination of the protein and

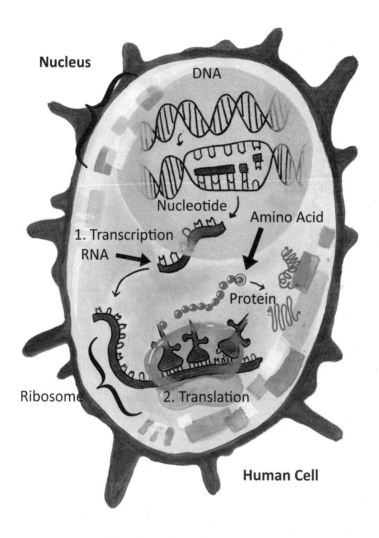

Figure 4. How a cell works. A simplified view of a 'standard' human cell. It shows how information gets from the DNA lending library to the protein factories (the ribosomes) via the messenger molecule RNA. This is important in the context of how our genes shape our cells and ultimately our bodies.

loss of function. Some of the most dramatic genetic diseases are caused by loss of function mutations, including Duchenne muscular dystrophy.* Small changes in gene sequence can lead to significant effects; the deletion of a single amino acid in the cystic fibrosis transmembrane conductance regulator (CFTR) gene is enough to break it and cause cystic fibrosis (CF). The most common CF mutation, called F508del, lacks a phenylalanine amino acid at position 508 (hence 'del' for 'deleted'); in a piece of good news a new drug, Trikafta, can reverse the effects of this mutation and treat this previously devastating condition.

Some mutations are neutral – encoding a like-for-like swap; some mutations (like deletions) are bad. But a smaller subset can improve the protein. Or at least provide a basis for future gains. Evolution acts upon these beneficial changes. A textbook example is the peppered moth, which has two variants, dark and light. During the industrial revolution, the dark moth began to predominate, because trees became caked in black soot, providing it with better camouflage; lighter moths did better in the cleaner countryside and thrived once air quality improved.

Each of our genes has multiple possible variants, called alleles. Different alleles can associate with different physical

---

* Duchenne muscular dystrophy (DMD) is caused by loss of function mutations in the dystrophin gene. Genes have a weird and wonderful array of naming conventions: sometimes the name indicates the function, other times they take their names from the disease they relate to and sometimes just through the whim of a fruit fly geneticist. DMD is named after the French neurologist Guillaume Duchenne, who differentiated between a true 'eye smile' and a false 'mouth only' smile. Even STOP codons can have funky names; they are called amber, ochre and opal – because the first discovered one (amber) made a yellow pigment.

phenotypes. Some have large effects, like the CF mutation, some have noticeable effects like eye colour, others have effects only in combination with other genes and some have no effect. Most changes between different alleles comprise a single nucleotide (one letter in the code): these are called single nucleotide polymorphisms or SNPs (pronounced snips).

SUCCESSION

Our genes, with all their wonderful variety, are packaged within the three billion base pairs of human DNA and organised into individual packets called chromosomes. Each cell contains forty-six of these packets, organised into twenty-three pairs. At various stages of the cell's life cycle and using a sufficiently powerful microscope, chromosomes can be seen in a distinctive off-centre X shape. Chromosomes vary in length from about 50,000 to 260,000 base pairs. Each of the matching chromosomes holds one of a pair of each gene. A major advantage of having two copies of each gene is redundancy: if one copy misfires the other can replace it. We get one half of the chromosome pair from each parent, they in turn got one of each pair from their parents. This means you share fifty percent of your DNA with each of your parents, twenty-five percent with each grandparent and so on. The jumbling up of genetic information at each generation is one of the major reasons why sexual reproduction evolved; it broadens the diversity upon which natural selection acts.

Our genetic relationship with our siblings is more complex than with our parents. Each sperm from your dad contains fifty percent of his chromosomes, each egg from your mum has fifty percent of hers. The same goes for your sisters and brothers, but you do not necessarily receive the same fifty percent. If you compare two different sperm from the same father (or eggs from the same mother), on average they will share fifty percent of their genes. So, on average siblings share fifty percent of their genes. But as pointed out to me by my colleague Dr Vanessa Sancho-Shimizu, the random selection of chromosomes can result in siblings ranging from zero to one hundred percent genetic overlap. Which means that theoretically, if two identical sperm fertilised two identical eggs, one hundred percent identical siblings could be born at separate times – asynchronous identical twins! The chances of this are vanishingly small (one in seventy billion). The maths works like picking red and green socks from each of twenty-three drawers; because we have twenty-three chromosome pairs, there are 8,324,608 possible combinations for each sperm or egg, and you multiply these enormous numbers together. Similarity increases the closer your parents are related. This explains why it isn't advisable to sleep with your cousins; the greater your proximity in the family tree, the more genes you will share, increasing the likelihood of any ill-begotten offspring acquiring two copies of any faulty ones. If one of your parents carries a faulty copy of a gene, the healthy copy from your other parent should take over in your chromosome and allow you to function normally. But if your parents are more closely related, they may both be carrying a faulty copy, potentially leaving you with a genetic disorder.

Intermarriage can be a challenge in isolated communities – as in European royalty in the nineteenth century, a lack of diversity increases the risk of offspring getting two copies of a mutant gene. It can also cause problems when a small number of individuals set up a community, leading to something called a founder effect. If a colony derives from a handful of pioneers, all of their descendants will draw gene variants from the same, smaller deck; if any of the founders carried an unusual mutation, it will have a much higher penetrance, posing a theoretical challenge to sustained life on Mars.

Small, interbred communities are often cursed with much higher levels of rare genetic diseases. Knowing the degree of consanguinity of everyone around you can reduce that risk. Iceland has a high degree of interrelation: it only has 375,000 inhabitants with relatively little influx, leading to quite a shallow pool from which suitors can fish. Fortunately the *Íslendingabók*, or 'Book of Icelanders', records all Icelandic genealogy (and by inference genetics), enabling you to see how closely related any future partner might be. The *Íslendingabók* has now been digitised into an app, which really should be called Kin-der, but I guess that doesn't work in Icelandic. Even in larger countries everyone is a bit related – on average in the UK people have 193,000 living cousins (up to sixth cousin). You could fill the town of Norwich with your relatives (and indeed some people do).

Regardless of the interrelatedness of our parents, genetics affects our health, but not in the simple way we might imagine (or want). Stories such as the 'Jim Twins' are alluring but not really representative of how genetics works. The story goes like this. In 1940, a pair of twins was separated at birth and adopted by different families. They both ended up being named Jim, both owned a dog called Toy, both married twice, first to Linda then to Betty, and they both named their first son Jim. Jim 1 and Jim 2 also both smoked and drove a Chevy. Sadly, this doesn't really show anything, except if you look hard enough, you will find coincidences.

A lot of American men in the 1940s were called Jim and a lot of women Betty, nearly everyone smoked, and Chevrolets were an extremely common car; the matching dog names is more striking, but you can bet the Jims read different newspapers, took their coffee in different ways and supported different baseball teams. Humans love finding patterns, it doesn't make them meaningful.

Simply put, human genetics isn't simple. Whilst each pea cell contains more DNA than a human one (4.5 billion bases of DNA compared to a human's 3 billion), one could argue that humans are more sophisticated. And if so inclined, you would almost certainly win that argument with a pea plant.* To understand more about human genetics, I spoke to Dr Adam Rutherford, geneticist, broadcaster and author of several books, including *Control* (which tells genetics' darker history).[1] When we met via video conference, we connected, as all right-minded people should, through a shared passion for cricket; having discussed the woeful form of the England ODI team we got down to genetics. His first point was that genetics is much, much more complex than the school textbook version. There are, of course, some important conditions associated with single genes – cystic fibrosis, sickle cell anaemia and types of haemophilia. But 'genetics is probabilistic rather than deterministic', he tells me, so most of our traits are not determined by simple Mendelian laws. Examples of single gene traits are almost uniformly trivial, such as earlobe shape, tongue curling and

---

* Though one could also suggest that if you find yourself resorting to arguments with pea plants, things aren't going all that well; still, pontificating with *P. sativum* beats some of the discussion on social media.

smelly asparagus wee. And asparagus wee is not even that simple, with 871 different gene mutations linked to asparagus anosmia (not being able to smell it).[2]

Even the classic example taught at schools – brown/blue eyes – lies. It is taught as monogenic with brown being dominant to blue. But it is actually driven by two different genes, OCA2 and HERC2.[3] The protein encoded by OCA2 affects the production of melanin: if you possess a working copy of it, your eyes contain brown pigment. However, you also need a working copy of the HERC2 protein to activate OCA2; therefore, you can have two 'brown' OCA2 copies and still have blue eyes, because you have two 'blue' copies of HERC2. And of course, that doesn't take into account green, hazel, light blue, dark blue, amber, grey or any of the other iris colour variants.

Whilst Mendelian inheritance of pea flowers can be understood relatively easily, it doesn't really explain the complexity of humanity, especially for traits such as intelligence – or longevity. Other non-Mendelian inheritance types include: incomplete dominance (where the phenotype blends two genes – like pink flowers, but don't tell Gregor), co-dominance (where both genes can be expressed – for example, both black feathers and white feathers on the same chicken), gene linkage (where two genes are so close together; phenotypes nearly always appear together – like blond hair and blue eyes) and multiple alleles (which lead to more than two possible outcomes).

To complicate matters further, we don't just get our genetic information from our nucleus! The mitochondria, the powerhouse of the cell, is actually a bacteria that cohabits happily with us. We inherit them exclusively from our mothers (an egg being ten million times larger than a sperm) and the DNA powering the mitochondria can be exclusively traced backwards to a single ancestor (called Eve).

Genetics interweaves with ancestry, but Rutherford stressed that even that relationship encompasses more than a simple parent-to-child narrative over the generations. 'Genetic genealogy and genealogy are not the same thing and they diverge over time'. While we get our genes from our parents, the process of halving and mixing genetic information in each generation means that by the time we get to our great grandparents, we only receive 12.5% from each. If you backtrack eleven generations, you can carry zero genetic information from half of your direct-line ancestors. I can trace my family tree back ten generations to John Tregoning born in Gwennap, Cornwall, sometime in the early 1600s. And whilst we share the same first name (as did five others of my direct-line ancestors, we Tregonings are nothing if not imaginative*), we probably share only our Y chromosomes; there is a strong chance one of my two sisters has no directly inherited genes from John Tregoning (the Gwennap one, not me – I know we are Cornish but come on).

I asked Rutherford about the larger interweaved family tree of humanity. He came back with tidings of great news – we all grow from royal roots: 'All Europeans are the direct descendants of King Charlemagne,' states Rutherford. However, this connection is pretty tenuous. 'All Europeans are also the descendant of Charlemagne's toilet cleaner and frankly anyone else alive at the time.' Rutherford uses Charlemagne as a benchmark, because a direct paper

---

* If you found yourself wondering, the word 'Tregoning' probably means 'of Connon's homestead' and stems from a small hill fort just outside Helston, Cornwall.

genealogy to the Holy Roman Emperor exists, and he lived far enough in the past (about one thousand years) that so much interleaving of human genetic trees has happened that everyone is related to everyone in the end. This mixing of lineages extends beyond European shores. The shared point of common ancestry for every person on earth is about six thousand years ago, which seems odd given the tribes of earth split far earlier than that. But as Rutherford tells me, 'If there's two things humanity is good at, it's moving and shagging.' Colonisation (and rape) led to the flow of European genes into other previously isolated family trees; which brings forward the common point of ancestry.

But what did my newfound knowledge of genetics tell me about my risk of disease and death?

RISKY BUSINESS

Time now for a critical concept: risk is relative. Unless you go to extremes, very few biological things are absolute. For nearly every situation, someone or something won't follow the pack (at least anecdotally). My maternal grandfather smoked, drank gin every day, fought in Burma (now Myanmar) in the Second World War and yet somehow survived to a ripe old age; Queen Elizabeth, the Queen Mother, survived to 101 in spite of a diet of dubonnet and gins, red wine and pink champagne. Everything we undertake has some form of risk; for any activity, no matter how saccharine, someone will have injured themselves doing it. The ninth-century Muslim historian al-Jāḥiẓ was crushed to death by his own library. And it's not just ancient scholars who fall foul of mundane things: in 2001, forty British people injured themselves on tea cosies, ninety-one with bread bins and six thousand while putting on their trousers; the 'good news' is that sponge and loofah injuries fell from 996 to 787.[4]

A key message to take from this book is that there is risk in everything you do or do not do. In my field of vaccinology, taking a vaccine carries a risk – adverse effects occur in a tiny percentage of vaccinated people; these are so rare they can be presented as the number of incidents per million vaccinations. BUT the risk of *not* getting vaccinated far, far outweighs this. During the COVID-19 pandemic, infection had a high risk of fatality, increasing from 0.2% in the younger groups to fifteen percent in the over-eighties. If you got vaccinated, the risk of death fell to nearly zero. Whilst some people experienced severe side effects, the risk of death from vaccination was effectively zero.[5] To help you visualise this imagine you are eighty and have to draw either a red or a blue pill from one of two tubs. The blue pills come from a pot labelled 'vaccine'; the pills in this pot come will give you a sore arm and a headache. The red pills from a pot labelled 'no vaccine'; they will give you no symptoms, but three out of the twenty pills in this pot will kill you. Would you still take the red pill?

CALCULATING RISK

Since most biological crystal ball gazing is quite murky, medical epidemiologists dedicate a lot of their time trying to link cause and effect. These causal linkages can be expressed as relative risk, calculated using the equation:

$$\text{Relative Risk} = \frac{\text{(Probability of event in exposed group)}}{\text{(Probability of event in not exposed group)}}$$

They then present the outcome from the risk equation as a multiplier; doing thing X makes you Y times more likely to get condition Z.

For a more concrete example than X, Y and Z, let's look at a causal risk that took a surprisingly long time to establish – smoking and lung cancer. Whilst German scientists collected some evidence connecting the two in the 1930s, it took large epidemiology studies in the 1950s to confirm the link. Richard Doll and Austin Bradford Hill led one of the most important studies. Hill was an ex-First World War fighter pilot, and all-round statistical wizard; he led the first randomised clinical trial ever, demonstrating that the antibiotic streptomycin could cure TB.[6] Before epidemiology, Doll served as a regimental medical officer in the British Expeditionary Force and found himself caught up in the retreat to Dunkirk.[7] To calculate the relative risk of tobacco and lung cancer, D&H quizzed forty thousand doctors on their smoking habits. They separated them into smokers and non-smokers and compared lung cancer incidence in the two groups. They published their results in their 1954 *BMJ* paper, 'The mortality of doctors in relation to their smoking habits',[8] which documented thirty-six deaths from lung cancer in the smokers and none in the non-smokers, with rates increasing in proportion to the amount of tobacco smoked.

Doll and Hill's study exemplifies how to connect a bad thing to a bad outcome; but plenty of badly performed studies have muddied the health-risk waters, either connecting good things to bad outcomes or bad things to good outcomes. To get accurate answers about the risk of an individual activity, other considerations need taking into account.

Hard outcomes. The more precise the outcome, the more precise the risk calculation. The question 'If I swing this axe at my arm, will it fall off?' is much easier to answer than 'If I eat twenty blueberries a day will I live longer?' The presence or absence of a limb being much easier to score than 'longer life'. In the case of Doll and Hill's studies, lung cancer was a clear, measurable outcome, and smoking a quantifiable input.

Confounders. Many other factors (called confounders) might contribute to lung cancer. The closer you can match your two groups, the more confidence with which you can draw your conclusion. This explains why Doll and Hill used doctors – they were relatively homogeneous in terms of social class and demographics (and, because it was the 1950s, sex).

Wisdom of crowds. To reduce the noise in biological studies, increase the number of people. The science term for people in a study is the n (for number) number, one of those everyday tautologies like free gift, naan bread, the hoi polloi, Mount Fujiyama, HIV virus, PIN number, D-Day and please RSVP.* Doll and Hill's study was large (n=40,000 doctors) and subsequently confirmed by multiple follow-ups. By way of contrast, my doing n=1 studies

---

* As a fun quiz – why are these tautologies?
Answers: gifts are by definition free; *naan* means 'bread'; *-yama* means 'mountain'; *hoi* translates to 'the', making 'the hoi polloi' 'the the many'; the V in HIV stands for 'virus'; the D for 'Day'; the N for 'number'; and SVP for *s'il vous plaît* – the French for 'please'. Brilliantly, there are one-word tautologies including ferryboat, pathway and, surprisingly, sledgehammer (sledge comes from *slægan*, 'to strike violently'). Thanks to Dr David Lowe (UCL) for those.

on myself is scientifically meaningless, but hopefully they serve an illustrative purpose.

Plausibility. Finally, the relationship needs scientific credibility. Smoking causing lung cancer is plausible because you inhale poison directly into your lungs.

RISKY INHERITANCE

Since everything has a risk, we can assign risk to our genetic makeup. But working out which genes cause diseases under what conditions presents enormous challenges. The strongest correlations are the easiest to solve – because the phenotype (that outward manifestation of our genes) presents so clearly: for example, the F508del mutation in the *CFTR* gene leading to cystic fibrosis (CF). This type of connection is called a single gene association or a monogenic trait.

But not all the genetic information resides in the bit of the gene that directly encodes the protein. The string of DNA letters that determine the string of amino acids in a protein is called the coding region. The coding region is extremely important, but the context of the gene in the wider chromosome plays a role, so mutations in these regions can also cause problems. Bits of DNA proximal to the gene are called non-coding regions and, rather than determining *what* the gene makes, influence *how much of it* is made. SNPs, those small differences in our genes, occur in both coding and non-coding regions of genes. For example, both F508del and rs62405860 are SNPs, the difference being that F508del defines a coding SNP in a known gene and rs62405860 indicates a non-coding SNP that apparently comes from Neanderthals and associates with hoarding.*

However, as we saw with eye colour, the impact of genotype on phenotype seldom is rarely as simple as one gene determining one feature. Traits determined by multiple genes are called polygenic. Genes need to be considered in the context of other genes: do the resultant proteins interact properly? Whilst genetics originally focused on the impact of individual genes, the complexity of the human condition requires a more holistic approach. This is where gene-wide associations studies (known more commonly as GWAS, pronounced as two words G-WAS, to rhyme with gas) come in. In these studies, researchers compare the DNA of huge cohorts of people with and without a certain condition. This may sound cripplingly expensive, but not all of the vast amount of DNA in a single human cell directly affects the proteins our cells make (somewhere between ninety and ninety-eight percent is non-coding or 'junk' DNA). To make life easier and studies cheaper, a slight shortcut is made in sequencing the DNA: rather than reading every letter, the common mutations – the SNPs – can be read. The level of statistical association with risk of disease is then calculated for each individual SNP and plotted in one of the more visually pleasing graphs in science – the Manhattan plot. This puts each gene along the X axis and the risk on the Y leading to little towers of genetic danger, like the Manhattan skyline. GWAS can be powerful and informative, but have limitations, centred on the probabilistic nature of genetics. At a population level they can

---

* Found in the regulatory region of *GGNBP1* for those of you really into your genetics.

indicate the risk of a condition; at an individual level they cannot.

The joys and challenges of science lie in its complexity. New technology peels back layers, revealing more, equally complex layers below. Genetic sequencing can now be performed on single human cells, revealing that our cells do not share the same DNA.[9] If you think about it at all, you probably imagine that we possess one set of genes reproduced faithfully in every single cell. In this version of us, if you took a cell from the top of your head and one from the tip of your toe, the DNA in their nuclei would be the same. In actual fact, our bodies are a mosaic of different mutants, a result of our cells taking different developmental pathways from the original zygote (the first cell that is definitively us, made from the fertilisation of an egg by a sperm). This heterogeneity will inevitably affect the risk of different diseases; if you get a mutation affecting a heart muscle protein but in a cell lineage that makes up only the eyes, the disease will not manifest.

PALM READING THROUGH GENETICS

Biology's complexity explains why I love research – pulling on the loose yarns that make up the cardigan of life. But science is endless and, sometimes, you need to use the information you have rather than dig deeper. In this case predict, to some degree, the risk of ageing diseases associated with certain genes (at a population level). The wealth of genetic information out there led me to hope that I could determine my own genetic risk score, since taking the family tree approach failed to provide any predictive insight. To do this, I needed to get my genes sequenced, the first of my experiments, and certainly the easiest.

But before I did that, I needed to consider the repercussions genetic testing could have upon my children. Sequencing differs from other diagnostic tests such as cholesterol measurement because of its broader familial impact. If my blood cholesterol is high it increases my risk of a heart attack, but it doesn't directly affect my offspring's health. Genetics tests differ because I have passed fifty percent of my genes to each child. My genes not only foretell my future health, but also have direct-line health reverberations for my children. If on screening I discovered I carried something terrible, there is a one in two chance my children would carry it too. This is particularly the case for something dominant and without treatment, like the gene for Huntington's disease – the irreversible neurodegenerative disorder. Knowing one's genes comes with a Cassandra-like quality; you may predict the future but will have little power to change it.

As a family we discussed the issue before I undertook the test. Having been given the green flag, I ordered the kit from 23andMe, the US-based genetic company. A few days later a tube arrived in the post into which I spat a sizeable volume of saliva and returned to sender. However, my first attempt failed – I had provided the wrong kind of spit, or there wasn't enough DNA. I have form on this – during the COVID pandemic my PCR swabs routinely failed because I failed to scrape enough nasal cells from my nose with a Q-Tip. Which goes to show, just because I can do science in a lab, I can't necessarily do it at home. It also raises the question of what else gets sent through the post in the name of science and medicine; and how posties should thank the inventors of the Jiffy Bag.

Second spit lucky, I got the results. The first level of information provided was somewhat trivial – and because they reported such visible phenotypes I could have done it more simply by looking in a mirror (and more accurately too). Genetics predicts likelihood rather than actual outcome, so it got a fair few wrong – for example according to my genes there is a seventy-eight percent chance my ring finger is the longest – where in actual fact my index finger is longer. The good news is that genetically speaking I am unlikely to have back hair or go bald. I also have wet earwax, which I can confirm by digital excavation, though it says nothing about the flavour (alas, not good). Best of all I share a protein, found in fast-twitch muscle fibres, with elite athletes. My elite athleticism is something I had always suspected – shame I never trained to realise my huge potential. The report also suggested a genetic basis to the fact that, unlike my wife, I don't get bitten by mosquitos, which does nothing to improve marital harmony. It said nothing about midges though, those devilish wee beasties!

In addition to the dampness of my ear wax, the genetic analysis contained information about both health and ancestry. The DNA database is now substantial – one source estimates that 26 million people had undertaken commercial genetic testing by 2020.[10] An unexpected consequence has been (re)uniting family members who did not know they were ever parted. But not every genetic reunification story ends happily. Many are heartbreaking family tragedies – people discovering their genetic father or mother were not who they thought they were, and in one case that two embryos were somehow swapped over during IVF and then born to the wrong mothers. Ancestry testing has also revealed more sinister stories, where individual sperm donors conceived scores of children. One Dutch fertility doctor, Jan Karbaat, is reported to have fathered at least forty-nine and up to two hundred of his patients' children.[11] Karbaat isn't the only serial donor: Donald Cline in the USA begat ninety-four of his patients' offspring. And another man from the Netherlands has allegedly sired more than five hundred children, making his own single-handed attempt to replicate the impact of Genghis Khan, whose Y chromosome can be found in 8% of Asian men and 0.5% of the global population.[12] I briefly engaged in the family tree aspect and discovered a third cousin in Australia called John Gurner, comfortingly unsurprising given my great-grandmother was born Shirley Gurner. I discovered no surprise half-siblings – reassuring.

As well as revealing missing cousins, ancestry DNA reports have been extended to mapping the geographic range of our ancestors. Rutherford explained the limitations of ancestry mapping using this approach; he emphasised that race does not have a genetic basis. The intermingling of genes basically makes 'race' meaningless. Furthermore, the way ancestry genetics tests assign origin is through similarity to people currently living in specific geographic regions, but that represents a snapshot of the current day and says nothing about where population groups came from. It wasn't a huge surprise that 99.9% of my genes overlap with other people from Britain and Ireland. Though approximately two percent of my DNA has Neanderthal origins. This apparently means I have a genetic tendency to hoard things (true), get lost easily (true), not fear heights (not true) and be a better sprinter (not true): in four fifty-fifty calls two of them were true – imagine the odds of that!

I thought the health information would be more relevant. But an important caveat was what it *couldn't* tell me. 23andMe received its license from the FDA which puts restrictions on what it can and cannot predict.[13] The sequencing approach they use also has limitations, meaning they cannot detect certain conditions.

Genetic testing can provide a probabilistic likelihood of a connection between certain SNPs and disease outcome at a population level. My having a variant in a specific gene doesn't tell me I *will* get a specific condition. But it does tell me that across the whole of humanity (that has been measured) there is a percentage increase in risk in people with that variant. It's the difference between saying if you live in Glasgow there is 2.5% chance you will become a heroin addict (which isn't true) and there are 2.5% of people in Glasgow who are problem drug users (which is). The test revealed that I don't carry risk variants for any of the forty-six traits they measure or any of the fourteen health predisposition variants covering a range of disease including breast cancer (*BRCA2*), Parkinson's disease and age-related macular degeneration. It was all quite boringly reassuring.*

One other huge consideration with 23andMe and other DTC (direct to consumer) genetic tests is ownership of the data. After some political infighting, the original Human Genome Project became freely available to everyone for perpetuity. DTC genomics testing works on another model. 23andMe was established as a commercial venture. They own an enormous database of human genetic variance; this can then be sold to drugs companies – for example GSK invested $300 million in 23andMe in 2018.[14] However, the data gathering process muddies the water: we the punters pay the companies for the privilege of collecting (and then selling) our most personal of data. It is not dissimilar to

---

* But does make it slightly harder to write a narrative about.

how big internet search companies work: we make a contract with them (knowingly or not) where they scrape our browsing data and sell it to others in return for being able to easily find a recipe for marmite pasta, whether farting burns calories and other such vital searches.

The history of getting people to pay for someone to harvest their health data goes back to the nineteenth century. One programme was run by Francis Galton, Darwin's cousin. I think it's not 'too woke' to say that Galton's legacy isn't great. He coined the word eugenics, an idea ultimately adopted by the Nazis. In 1884, Galton set up an anthropometric laboratory at the International Health Exhibition, held in South Kensington, London, on the site of the future Imperial College London. The Victorians paid three shillings to have a range of values measured, for example punching strength (a vital measurement we don't collect enough data about these days). At the end, the participant received a souvenir capturing their vital statistics, but Galton got a huge dataset upon which he could perform further analysis – and gain financially. Unlike Galton's punch-drunk punters, the ability to opt in or out of further analysis is much easier for modern day data donors. It comes down to the flick of a virtual switch: data from 2021 suggests that eighty percent of participants opted into the wider gene pooling. Which gives you a choice – with a rose-tinted mindset it could be said that the data curation can benefit humanity through the development of new drugs; with a more cynical view you could argue that you are paying money to enable the sequencing companies to get rich off your DNA. And of course, once the data is out there on a database, it is available for someone to steal it, as seems to have happened; another reason not to turn on the DNA relatives feature.[15] Whilst also not without risk of hacking, it is probably better to give up your DNA sequence to a public project – such as the NHS-sponsored Our Future Health, where the data will be freely available to all researchers everywhere in the spirit of the original genome sequencing project.

Whilst discovering I had a slight genetic tendency to dislike coriander satisfied my whimsy, the test didn't change my thinking about my upcoming deterioration and death. Individual genes are not deterministic, at best indicating a level of risk. There is an adage attributed to various people, but most likely coined by the nutritionist Judith Stern: 'Genetics load the gun, but the environment pulls the trigger.'[16] Stern means that genes only form part of the picture; whilst they predispose us to disease, it's what we do to our bodies that shapes our outcome. Since genetic testing failed to predict ways to stave off *future* unhappiness, except maybe to avoid carrot and coriander soup, I decided to take the next step – measure (as much as I can) what is going on in my body right now as a way to predict what might go wrong and how to avert it.

CHAPTER 3

A Picture of Health:
Diagnosis and Prognosis

'Diagnosis is not the end, but the beginning of practice.'

Martin Fischer

Trying to decipher what is going on under the skin to extend life has a long – often misguided – history, from palm reading, through urine swilling to bloodletting, phrenology and astrology. Diagnosis, the word for this process, derives from the Greek for 'to discern'. Having discovered relatively little about myself through genetic testing, I turned instead to a full body diagnostic check-up. Medical doctors use diagnosis to identify what might be wrong with you, so they can make you better. As medical science progresses, more sophisticated tests are used to determine treatments for more complex conditions.

In emergency medicine, diagnostics tells clinicians what to fix and what order to fix it in, enabling them to save life, there and then. But in the context of ageing and disease, the longer-term prognostic value of diagnosis matters. It can take many forms. Some diagnoses follow physical investigations – a bone poking out of the skin suggests it may well be broken! Other measurements directly relate to the function (or dysfunction) of a particular organ. But diagnoses can also be based on correlates, measurable markers in some way related to disease, like TNF in the tomato study. Chemical tests performed in the lab are grouped under the catch-all of *in vitro* diagnostics. Whilst other bodily fluids are used for specific indications, for example urine for kidney function and nasal swabs for respiratory infections, most diagnostics concentrate on the blood.

BLOODY RICH

Having a full-body MOT did not come cheap; I didn't pocket much change from £200. There is a lot of money in blood. The total global value of the diagnostics market in 2023 was $99 billion;* not all of that came from predicting diseases of ageing, a large chunk derived from COVID-19 testing.[1] The money to be made in diagnostics has attracted some bad players, most notably Theranos, which falsely claimed the ability to read a huge range of indicators in a single drop of blood. The charismatic CEO of Theranos, Elizabeth Holmes, finds herself serving an eleven-year sentence for wire fraud, joining a bunch of other tech gurus behind bars or facing jail. Overhyped Silicon Valley sales pitches aside, our blood, with its cargo of cells and chemicals, contains a wealth of information. Because it bathes all of our organs, blood gives us a window into our body's health.

---

* $100 billion sounds a lot, but worth noting Amazon is worth one thousand times more. We are apparently more interested in getting ephemera instantly shipped to our door than learning about our health.

Blood testing began in earnest in the 1900s, building on the work of Karl Landsteiner, who identified the different blood groups (A, B, O). One of the earliest proponents of diagnostic blood testing was Sir William Osler, who shaped medical training as we know it today, emphasising the need for bedside instruction. He is reported as quite the wag, and one anecdote relates how he used diabetic urine to test student observation. He appeared to dip his finger in the urine and taste it, then encouraged the students to follow suit, which surprisingly they did and unsurprisingly found revolting. The joke somehow being on them – because the 'observation' part of the test was that he dipped his *middle* finger into the pee but then licked his *index* finger; ho, ho. When not bullying students, Osler developed a range of tests for estimating levels of haemoglobin, cells or bacteria in blood. The automation of blood testing began in the 1950s when Leonard Skeggs, a naval Second World War veteran and survivor of the sinking of USS Hovey in the Philippines, developed a process called continuous flow analysis, enabling sequential running of different tests. It was a huge breakthrough – being able to run one test a minute! Since then, the scale of testing increased dramatically, with UK hospitals running more than one billion tests a year at a rate of 300,000 per day (fifteen per person per year).

I opted for the full works package, which was how I found myself wandering along Kensington High Street first thing on a Tuesday morning. Walking down a street devoted to health and wealth, I found the blood test clinic squeezed between a high-end bakery selling croissants at £10 a pop and a 'wellness' centre offering bespoke nutritional advice on how to counteract the negative impact of expensive buttery pastry. Having been turned away for arriving too early, and after spending twenty minutes catching up on emails on a church bench (ruing my lack of croissant), I eventually crossed the threshold. A Polish nurse welcomed me for the first of her eighteen appointments that day: a relentless stream of affluent middle-aged people worrying about their health. Before taking my blood there came a number of less invasive tests used to evaluate health. The simplest of which were height and weight. This caught me off guard and I didn't pull myself up to my full height (the consequences of which will become clear later).

We moved onto my nemesis – blood pressure measurement. Which turned out to be higher than it ought to be on the first attempt. I could look at this in two ways: either my blood pressure is too high (which I am not prepared to admit) or I get the dreaded White Coat Syndrome. Blood pressure responds to stress; visiting a GP surgery is not the most relaxing of occasions. Waiting impatiently for your appointment in a tired reception area, surrounded by other health-anxious people, with nothing but five-year-old copies of *Woman's Weekly* with the recipes torn out, is the opposite of a Zen rock garden. Which means by the time you get your blood pressure taken, it may already be elevated. Worrying about the results can further raise the mercury; having high blood pressure can indicate other underlying conditions, which causes one to worry, lifting it higher. When I visit the doctor's, my BP often drifts across the line in the wrong direction. To get around White Coat Syndrome, you can get home blood pressure units. In my experience, these aren't necessarily any better. Though again, I may not have applied ideal test conditions, doing it at my stepmother-in-law's house with the whole family watching and the children fighting over whose turn it was next. The whir of the blood cuff tightening triggers a Pavlovian reflex in me, sending my blood pressure skywards. Finally, after a few stern words with myself and a deep intake of breath, it returned to the green zone (on the third measurement) and we moved on.

The next part wasn't much of an improvement – blood-letting or 'phlebology', which sounds better until you discover it derives from the Greek for 'vein-cutting'. The nurse took six vials in total: four yellow-capped, one purple, one green, each of which holds about ten millilitres. As the nurse inserted the needle, I mentioned to her that I didn't enjoy giving blood, to which she deadpanned: 'No dear, no one likes giving blood.' As with much of the kit used and disposed of daily in the health services, blood tubes are elegantly designed. They are also known as vacutainers because they contain a precisely calibrated level of vacuum to draw the exact amount of blood required (-75 mm Hg for a ten-millilitre blood draw). The cap colours indicate what else the vial contains, as different tests need the blood to be in different states. Whilst the purple and green tubes contain anti-coagulants, which stop the blood clotting, the yellow tubes actively encourage clotting, freeing up the liquid part (plasma). Most of the tests are performed on the liquid part, hence a greater proportion of my precious life fluid went into gold-capped tubes. Even the needles used are a relatively modern invention. Christopher Wren, he of St Paul's and Pembroke College Chapel (Cambridge) fame, developed the first injection technology, but surprisingly using an animal bladder and a goose quill to deliver drugs wasn't that effective. A drawn metal needle was first used in 1844, but disposable needles were only developed to deliver individual morphine syrettes during the Second World War. All of this technology goes straight into the sharps bin after use. The NHS in the UK disposes of 133,000 tonnes of plastic each year, 2.6% of the entire country's five million-tonne total of plastic waste. Reducing this presents huge challenges, as most of the plastic is deliberately single-use to stop cross contamination; we have the same problem in labs – for all my careful sorting of milk bottles at home, I can generate an entire bag's worth of plastic waste a day at work that goes straight into the incinerator.

Having got my brave boy plaster for giving blood, we moved on to the other sample – urine. This presented the usual challenge of trying to fit three hundred millilitres of pee into a twenty-millilitre pot. I deposited the slightly damp tube in an anonymised locker, made up the usual lies about how much alcohol I drank each week and left.

What did they measure in my samples?

TESTING TIMES

Tests work by quantifying chemical components within a sample. Some tests directly measure the levels of these components. But others, known as biosensors, utilise biological reactions to infer their levels from the amount of reactivity. For example, home-use glucose strips contain an enzyme called glucose oxidase which reacts with the glucose in a sample of blood. The reaction releases electrons, and so the strength of the resulting electrical current signifies the level of glucose. Modern versions of this test use as little as one microlitre of blood, a tiny pinprick. An alternative biology-based approach is the immunoassay. These tests take advantage of the incredible specificity of a biological molecule, the antibody. As part of the immune response, antibodies recognise structural motifs in other biological molecules, such as those derived from bacteria or viruses. We can repurpose antibodies to help us detect a huge range of biochemicals. As such, they form the backbone of a huge variety of tests, from the dreaded COVID lateral flow test to the pregnancy pee stick.

MEASUREMENT FOR MEASUREMENT

A vast range of things can be quantified in the human body; in my deluxe package I had 150 different tests performed. The results came back as a string of numbers: for example, my blood contained 5.4 mmol/l glucose, which is not very helpful in and of itself. To contextualise them, results are presented in relation to a range of values: for non-fasting blood glucose the range goes from low (<4) through optimal to high (>5.6). These ranges are determined from huge population-level studies. Which means it's time for a quick lesson in clinical medicine and statistics:

1. Study design matters. Some studies look forward, called *prospective*. Prospective studies take a group of people as similar as possible to one another but differing in one key variable – in this case blood glucose – and measure the likelihood of disease: type 2 diabetes. Other studies look backwards, called *retrospective*. These studies flip the perspective, taking samples from

people with and without diabetes to compare their blood glucose.

2. Bigger is better (always). It is impossible to measure everything in everyone; therefore, a sample of the whole population is measured. But the more people the better: same as risk, same as clinical trials.

3. Meta is best. Combining multiple studies gives more power as it will exclude some of the bias. The guidance used for glucose levels and diabetes comes from 102 other studies, combining the data from 698,782 people.[2]

4. Studies can be (and often are) wrong. Scientific knowledge constantly evolves, the way we look at the world changing as our understanding grows. In the case of disease markers, one problem historically has been an overrepresentation of white European men in studies; what is true for them may not be true for other demographics. This bias has led to a range of interventions that protect men but fail women, from stab vests to crash test dummies.[3] The 'reproducibility crisis' also contributes to uncertainty; if performed properly and methodically, a study should be reproducible. Sadly, for a number of reasons, ranging from fraud, through incompetence to selective reporting, many unrepeatable studies get published. Hence looking at meta-studies is preferred.

5. Tending to the mean. People tend to cluster around a median value, giving a bell-shaped *normal* distribution. But there will be noise, and healthy ranges are based on a statistical value called the standard deviation (or SD). I'm not going to explain it here, on account of not really knowing how it works, and just

use the 'apply standard deviation' function on my stats package.*

6. You can be a healthy outlier. In a normally distributed population, sixty-eight percent of people fall within one SD and ninety-five percent of people fall within two SD of the mean. But statistically that means there are outliers, who exist outside the normal range and yet remain healthy. The more tests you perform, the more likely the chance of discovering outliers. It's a bit like flipping coins: do it enough times and you will get 'impossible' sequences, like Derren Brown's flipping ten heads in a row, which actually took him nine hours, rather than the one minute shown in his video.

PREVENTION IS BETTER THAN CURE

Why undertake all this effort? Just because we can measure 150 things about our bodies doesn't necessarily mean we should. The key purpose of diagnostic tests is detection and prevention. For any health condition, earlier detection improves the likelihood of resolving it or, even better,

---

* Of course, I know how it works really, I just wanted to be cool and pretend I don't do maths. The standard deviation is the root of the sum of all the differences from the mean divided by the total number of things being tested. If your sample is highly variable, the standard deviation will be large as a proportion of the mean. For example, the standard deviation of height in Tolkien's fellowship is 30 cm (nineteen percent of the mean: 153 cm), because five of the heroes were very small and four unusually tall. However, if you take just the hobbits, the standard deviation is 1.5 cm (one percent of the mean of 125 cm). Things that cluster closer together have a smaller standard deviation. These measurements are pre-Ent-draught; I wouldn't want to be accused of inaccuracy in my stats relating to fictional beings.

preventing it from becoming serious in the first place. For example, type 2 diabetes is not a binary (ill or not) condition: the higher the glucose rises, the worse the disease gets; importantly, a pre-diabetic state exists where blood glucose is higher than normal but not yet in the dangerous range. Early diagnosis of elevated glucose gives more time for behavioural adaptations and treatment to have an impact before damage occurs.

Whilst testing 150 markers is a bit OTT, testing a much smaller number of well-established risk factors can benefit the nation's health enormously. In the UK, the NHS offers a screening programme called Health Check to people over forty years old, every five years. It includes some basic tests (blood pressure, blood sugar, cholesterol, height, weight) that predict the most likely causes of disease: heart attack, strokes and diabetes. In the first five years of running, the programme prevented 2,500 cardiac events through a combination of behavioural advice and drug prescription.

Why don't people take advantage of the free testing? Between 2012 and 2017, 9.6 million people were offered a test, but only 5.1 million took it up.[4] Attendance increased with age, presumably because death's long shadow looms larger as our years increase. Women are also more likely to get checks than men. This does double damage, as men have a higher risk of cardiovascular disease. The lack of male uptake conceals a lot of psychology, loosely summarised as 'men are idiots'. For example, one survey suggested that forty percent of men prioritise their pet's health over their own. In another survey, sixty-five percent of men believed themselves to be healthier than others, which like surveys about driving* shows a

---

* Ninety-three percent of Americans rank themselves as better-than-average drivers.

blindness to one's own fallibility – there can only be fifty percent of people better than average. Of course, this 'better than average' effect doesn't apply to me ... I *know* I am smarter, taller and funnier than average. And testing can be scary, not in the 'I'm afraid of needles' way, but in the 'I'm afraid of the implications of the result' way. As in the old (but in this case wrong) adage, what you don't know can't hurt you!

The NHS programme is highly effective because it directly links the test and the healthcare provider. When GP tests identify something wrong, they can prescribe medicine to fix it. Tests through the NHS also often come with a consultation,[*] especially if the results may have consequences. Private providers give you a snapshot of health, but beyond telling you 'eat less, do more' often cannot prescribe further treatment. If anything is flagged as being really serious in a private test, the next step is often 'go back to old GP', which can put an extra burden on state healthcare.[5] Anecdotally, I have seen this to be the case. A friend discovered unusual liver readings after a commercial test, triggering a series of further, more expensive tests (done by the state health system) which all came back negative.

The small battery of tests the NHS offers is cost-effective and can have a big impact. Counter-intuitively, having more tests might actually tell you less! Coming back to standard deviation and the calculation of normal ranges, the more tests performed, the more likely a false outlier; essentially, for every twenty tests, one could be misleading. The measurements themselves are right (at that moment in time), but the interpretation of what they mean might not be. Since I

---

[*] Private tests also offer consultations, I skipped mine because I didn't want to hear any bad news.

had 150 things measured, five to ten of the tests could fall outside the reference range and yet not be a problem. On this basis I chose to ignore my low free androgen index – because this apparently meant I have fertility problems (my two children being reasonable evidence to the contrary).

RESULTS DAY

A few days after my blood test, I received the email containing my results. The first disappointment: I was shorter than I thought. To the delight of my son (and my friend Lucy), the report claimed my height as 181 cm, just short of the six foot I had convinced myself I was. I either have shrunk or didn't stand up properly at the test (or, whisper it, never reached six foot). Likewise, at this clinic I weighed 78.5 kg, but within a two-month period I fluctuated from a high of 83 kg (at my in-laws') to a low of 75 kg (at my parents'). Which goes to show how variable even a 'simple' measurement can be. It also showed the ease of cherry picking values when it comes to your own data.

Having got over the shock news that I would never play for the '95 Chicago Bulls, I turned to the rest of my results. Some of the data was reported as a direct marker of function – cholesterol for heart disease, glucose for diabetes, red blood cell count for anaemia. Other data were indirect markers of function, for example cystatin C is generally harmless, but high levels in your urine indicate problems with the kidneys. Reassuringly, my Cystatin C and therefore kidney function were fine (not that I had any reason to think they might not be, except for the occasional bright yellow piss). In other great news, my metabolic age matched a much younger man, well somewhat younger, OK five years younger.

Apart from my youthful metabolism, what did the tests tell me? One of the main outliers with a big red flag was C-reactive protein, commonly known as CRP. CRP is widely used as a marker of inflammation; the liver produces it in response to infection. The sky-high level initially caused me alarm, until I remembered that on the day of the test, I had a streaming cold. All told, quite an expensive way of being told my nose was running.

But the report contained 148 other results to pick through; since my wife had also had her bloods done, we decided to make a fun game of comparing results.* Reassuringly, the majority of our results fell in the green zone. Even my blood sugar came back fine – in spite of my habit of putting sugar in tea. However, we eventually reached some results that I couldn't just explain away as easily as CRP or free androgen. Firstly, whilst my BMI (body mass index) fell in range, I found myself near the top end of normal and my tummy was a bit larger than it should be. Admittedly these didn't need a battery of tests, just a mirror, a tape measure and some scales, but still in the context of my ageing, a bit of a wake-up. Secondly, my blood pressure, even after entering a Zen-like trance, registered on the high side and coupled with high levels of cholesterol, especially LDL (the bad one), did put me in a slightly risky category for strokes and heart attacks.

UNIQUE SNOWFLAKE

Discovering I was a bit fat with a slightly elevated risk of a heart attack (the number one cause of death in men my age – see Table 1) felt a pretty poor return on investment. Why,

* And people say scientists are dull.

after my battery of tests, genetic and phenotypic, did I not find myself any wiser about my risk of dying, or even which parts of my body were ageing? The main answer is that I am a unique and beautiful snowflake, as are we all. Each of us has a single, never-to-be-repeated combination of genes. To quote Nietzsche as a hat tip to my angst-ridden teenage self:

> At bottom every man knows well enough that he is a unique being, only once on this earth; and by no extraordinary chance will such a marvellously picturesque piece of diversity in unity as he is, ever be put together a second time.

But at risk of contradicting the Überman himself, it isn't just how we are assembled, it is how we treat the edifice once built. In particular, our environment defines our health. In this case, the environment means not just where we live but all the things that we do to ourselves or that get done to us. This is sometimes called lifestyle, but that word implies choice: many exposures fall beyond our control, particularly at the interface of health and wealth. To understand disease risk better, we need to know the impact of all the things to which we have exposed ourselves. The sum total of our exposures is sometimes called the exposome. For complete understanding, we need lifelong exposure information, detailing everything we have ever eaten, breathed and touched as well as every ray of sunlight that hit our bodies, every cosmic ray that passed through us and every virus that ever infected our cells. We then need a mathematical framework to contextualise their impact.

Unfortunately, many of the statistical tests we use in medicine were developed to deal with much simpler data sets. The statistical test that I and my colleagues use most often is called Student's t test. It was not developed to test multifactorial biological risk; in fact, it wasn't even invented by someone called Student. A brewer called William Sealy Gosset invented it to compare batches of Guinness;[6] he published under the name Student to avoid others discovering trade secrets about the Black Stuff. Another important statistical measure – the median – was developed by Francis Galton to guess the weight of a cow at a county fair by finding the middle ground among all the other participants' guesses.[7] [8] Meanwhile, the exact test (an important measure of statistical significance) was developed by Ronald Fisher to see whether one of his co-workers really could tell if her daily cuppa was prepared by pouring the milk or the tea first.[*]

To better understand risk, I spoke to Sir David Spiegelhalter, Chair of the Winton Centre for Risk and Evidence Communication, University of Cambridge, after a talk at work in which he explained how statistics helped unmask Harold Shipman.[†] On the stairs down from the talk, I asked whether there was a way, given the enormous complexities of our lives, to start personalising our risks. He answered that rather than thinking about personalised approaches, we need to consider stratifying our risk into as large cohorts as possible. A bit like the strategy for Guess Who, where you eliminate people based on shared characteristics, but with fewer hats.

[*] Reader, she could.
[†] It came down to the timing of the deaths – Shipman's patients' deaths spiked during his afternoon home rounds when he visited vulnerable patients on their own.

To illustrate the point, Figure 5A shows one hundred mini-mes to represent the UK population by percentage. At this level, we can predict that everyone will die and make some generalisations about how, but not with much precision.

Figure 5A. If all the world was me. One hundred mini-mes to represent the whole population of the UK.

The first grouping is then to split by sex – putting me in a group of fifty. This gives us a bit more predictive power, and we can start to pin down a rough life expectancy and, based on previous data, some of the top causes of death.

Figure 5B. Cohorting people. I am in the fifty percent of the population that is male.

We then group by where I live – giving me a cohort of about five. This begins to match a bit more for other factors such as demographics and economics. Narrowing things down further.

Figure 5C. Subdividing the world into smaller groups. Within the fifty percent of population that are male, a smaller subset are the same age as me. You can subdivide further. This increases accuracy, but loses predictive power because we are all individuals.

We could further subdivide those five into smaller and smaller cohorts, but for each subdivision, we trade-off between greater specificity but less predictive power. There is a balance: as individuals we don't necessarily behave like our peers, but it makes exact prediction impossible.

GET TO THE HEART OF THINGS

I had set out to find out whether I could focus in on specific areas of my health and predict my death. I wasn't particularly successful in narrowing things down and, as I was about to find out first-hand, it is easy to get stuff measured but it is much harder to do anything about it. For me, blood pressure and cholesterol stood out as clear outliers. If I then return to my stratified risk cohort – the five percent of the UK population who are men aged between forty and fifty – the most common cause of death is heart attack. Health is multifactorial and each of the most common causes of death shares overlapping risks, but one line stood out in my personalised health report – that whilst my pulse remained low, reflecting my elite athlete status,* they recommended that I discuss the result with a doctor and considered having an ECG. So, with a heavy heart I headed off to my nearest cardiologist to find out what was going on and whether I would join one-third of my peer group and die of a dicky ticker.

* My interpretation, it didn't actually use the words 'elite' or 'athlete'.

CHAPTER 4

The Heart of the Matter:
Killer Number 1 – Heart Disease

'For where your treasure is, there will your heart be also.'
The Holy Bible: King James Version, Matt. 6:21

The most likely thing to kill me is my heart, and not in a soppy, poetic way, but in a large, vital muscle in my chest stops working, stopping the flow of blood to my brain kind of way. And whilst I am most interested in extending *my* own life, stats being what they are, *you* my dear reader are equally likely to suffer a massive coronary as me, so it behoves us all to get to know our hearts.

Let's start with an experiment. Get a tennis ball and squeeze it in your hand sixty times in a minute. Tired yet? Now repeat it, but twice as fast.

I imagine your arm muscle aches by now. Could you do it a third time even faster if I told you your life depended upon it? Well, there is a muscle in your body that contracts at least 60 times a minute and can go as fast as 220 times a minute and your life literally depends upon it. Can you guess what it is yet? It's the heart, of course.

On average your heart beats 100,000 times a day, approximately three billion times in a lifetime. The total number of times your heart beats in the span of your existence will be shaped by your overall level of activity and resting heart rate. Resting heart rate differs between different people; factors affecting the resting rate should not be that surprising – age, fitness, emotions, obesity, cigarettes and alcohol (the last three as we will serve as repeat offenders for all kinds of poor life outcomes). Resting heart rate directly correlates with disease; if it exceeds one hundred, your risk of a heart attack increases 1.5-fold.[1] It's easy to check – count the beats in ten seconds, if this exceeds seventeen, visit the doctor.

Smugly, I thought this would be a good place to show off my low resting heart rate, demonstrating how fit and healthy I am. The reason for this misplaced confidence was that my heart rate used to be around fifty beats per minute, which to my mind meant I could be considered a paragon of health. I've had a vague obsession with my heart rate since learning about Miguel Induráin, the five times Tour de France winner, as a teenager (much to my relief Induráin isn't one of the cyclists stripped of their titles for doping*). At his peak, Induráin had an insanely low resting pulse of twenty-eight beats per minute. In 2012, fourteen years after retiring and in the same mid-forties age bracket I find myself, researchers reassessed Induráin's physiology.[2] It remained off-the-scale good, his readings exceptional for someone half his age.

They say pride comes before a fall. And sadly, my pride in my heart rate was based on historical data, rather than

---

* Endurance cyclists, a bit like TV personalities and radio DJs from the 1970s, are a category of people whom you assume to now be disgraced, also making you question whether it is OK to use their catchphrases.

an accurate measurement. To correct the record, I first undertook the tried and tested approach of counting my pulse for ten seconds and multiplying by six; this did not give me the answer I wanted. Undeterred, I tried my daughter's Fitbit; this also failed to give me the desired answer. Then, in a particularly unscientific 'third time lucky' way, I used the pulse oximeter that we bought in a panic during COVID, which had remained in a drawer ever since. My initial attempt recorded no pulse at all, which was even more alarming. After I changed the batteries and scraped the corrosion off the terminals I tried again. My pulse remained stubbornly higher than anticipated. Sadly, all the signs pointed to my resting heart rate being sixty-five beats per minute. Which isn't too shabby (for my age) but not the magic sub-sixty number hoped for. It means my heart beats about ninety thousand times a day. It was an important scientific lesson: things change.

In spite of all this beating, the average human heart is surprisingly small, weighing about 290 g, or half a pound – slightly more than a quarter of a bag of sugar.* That places it as a somewhat mid-table heart. The fairy fly has the smallest heart of any living creature – and given a fairy fly only measures 0.2 mm long, its heart is genuinely tiny. The mammal with the smallest heart is the Etruscan tree shrew, coming in at about 0.02 g; remarkably it beats 1,200

---

* There are two types of scientific unit. Preferred in laboratories is the international system of units (known as SI units), including the metre, the kilogram and the candela (the unit of light in case you didn't know). But there is also the (much less formal) comparison used in communicating science: the SCI unit. For area it goes football pitch, then Wales. For volume: thimble, teacup, double-decker bus, swimming pool, etc. These units are more widely debated.

times a minute. The blue whale sits at the other end of the spectrum, with a monster heart weighing nine hundred kilograms – the same as a Mini car or an entire class of year four pupils, plus their teacher. A whale heart beats about six times a minute, moving 220 litres of blood each time. Even though the human heart is an Everton FC in terms of size,[*] it still shifts about one hundred millilitres a pump, working out as five litres (ten pints) a minute, the entire volume of blood in your body. A commonly held misconception purports that all animals have the same lifetime total of heartbeats, so the faster the heartbeat, the shorter the life; bats prove the lie, their hearts beat over one thousand times a minute and yet they live long lives, some species well over thirty years. Over a lifetime the human heart pumps about 350,000 litres, enough to fill the Serpentine in Hyde Park, London, twice.[†]

Fortunately, unlike choosing to squeeze a ball sixty times a minute, we don't have to think to make our heart beat. The heart possesses its own self-stimulating circuitry, called the pacemaker (more technically the sinoatrial node). Martin Flack discovered the sinoatrial node in a satisfactorily eccentric way by experimenting on mole hearts; history doesn't record from whence the moles came, this was presumably not the end of a wonderful day messing about in boats they had hoped for. The pacemaker beats of its own accord but is responsive to the body's oxygen requirements; when we require more oxygen it speeds up. A key

---

[*] Classic mid-table club. This statement may be triggering for some, but they have made poor life choices.
[†] For those of you not familiar with it, it is a large-ish pond, full of goose poo and pedalos.

time that we need more oxygen is during exercise. Whilst exercise is an important but voluntary activity for humans in the modern world, once upon a time the ability to run or fight was essential to survive as a species. To run faster or fight harder, the muscles need more oxygen – and to get more oxygen they need more blood pumped through them.

To prepare for these encounters, our bodies produce a hormone called adrenaline – often referred to as the fright, flight or fight hormone. Adrenaline peps you up, increasing blood flow, releasing more sugar into the blood and opening the airways. Because of its wide-reaching effects, adrenaline can be used in a range of emergency medicines. For example, the EpiPen, the emergency treatment for anaphylaxis (and rescuing Mia Wallace), takes its name from the less satisfying American word for adrenaline – epinephrine.

Cortisol, the longer-acting cousin of adrenaline, also plays a critical role in shaping our heart's response to the outside world. Whilst in the short term it performs a similar function to adrenaline, cortisol also plays an important housekeeping function. It wakes us up in the morning, when its release increases metabolism, firing up the engines for the day. Cortisol and adrenaline are grouped together as the sympathetic nervous system. The opposite system, the parasympathetic nervous system, supports resting, digesting and recovery. You need the up and at 'em stimulus the sympathetic nervous system provides, but you can have too much of a good thing: sustained sympathetic activation, particularly through cortisol, leads to long-term detrimental effects.

Whilst we no longer find ourselves chased by sabre tooth tigers, our sympathetic system still twitches in response to external stimuli. Let's perform another 'fun' thought experiment. Think back to your worst day at school. Maybe that time you forgot your gym shorts, or when you kicked the football through the headmaster's window or were caught chasing a duck with a tennis ball. Then remember how you felt: heart racing, skin red, potentially a loosening of the bowels. This is an example of psychology acting on physiology. These feelings are all driven by the sympathetic nervous system preparing you to run away; the teacher or the spectre of public shame replacing the sabre tooth tiger. These effects can loosely be grouped as stress. Luckily, in our thought experiment, the teacher forgives you and the stress goes away. Problems arise when stress persists, leading to unending overstimulation of the system. Stress and extremely heightened emotions can cause lasting damage to the heart. Most of the damage occurs slowly, but hearts can undergo acute emotional damage. When *Takotsubo* or broken heart syndrome occurs, extreme emotion causes one of the heart's chambers to stretch out of shape so that it resembles a Japanese octopus pot.

THE LANGUAGE OF THE HEART

In addition to hormones, the heart is also under nervous control. Nerves connect the heart to the medulla, which first evolved in fish and reptiles 500 million years before Lucy first set foot on the Ethiopian plains. Located in the brainstem, the medulla acts as a control centre for all the stuff that your body does without your conscious control (involuntary functions), including vomiting, sneezing and breathing. While you can't lower your heart rate directly, you can trick your lizard brain by breathing more deeply and quickly. The medulla recognises that the blood is more heavily oxygenated and so tells the heart that it doesn't need to move as much of it. Dizziness can follow.

Meanwhile, if the medulla detects a higher demand in the body for oxygen, it will tell the heart to beat faster. The responsiveness is incredible. Try for yourself. Count your heart rate whilst sitting down, now do twenty star jumps (officially the world's worst exercise*) and count it again; then wait one minute and do a final count. Mine went seventy, one hundred, sixty-five. The nerves transmit messages to the heart via electrical currents. As Frankenstein realised when he reanimated his monster from the dead, the body is awash with electricity. One of the first people to explore the interaction of electricity and muscles was an Italian surgeon – Luigi Galvani. He showed that connecting a frog leg to an electric current caused it to twitch. An experiment I replicated as part of my undergraduate studies, getting a dead frog to wave to a classmate – just as they mentioned some reservations about using animals for experiments. Suffice it to say, they were galvanised to change groups the next week.

The heart crackles with electricity and, to understand more about it, I reached out to Dr Rachel Bastiaenen, a leading cardiologist. One of the last times we had spoken was at her wedding, where her husband taught my then five-year-old son how to do handstands. I headed to South London to have a heart-to-heart with Rachel. We met in the surprisingly pleasant hospital lobby. Not really understanding how medicine works, and basing my expectations on *Grey's*

---

* One exercise tops this, the 'Oh-No', an invention of some evil moustachioed Sergeant Major at the Royal Military Academy Sandhurst. It involves a squat thrust and a star jump with you shouting 'Oh-No' at the jump's peak.

Anatomy and *Casualty*, I imagined Rachel arriving in blue scrubs toting a stethoscope; instead she was in business casual with ankle boots. It's probably fair to say Rachel isn't the tallest person I know, and is soft spoken, but her unassuming presence belies her stature in the field of the genetics of heart disease, having published more than fifty articles. Having quickly caught up on family gossip, we headed up to her office, where the first piece of good news was that when measured on a proper hospital machine my heart rate came in at less than sixty. Pride restored. But boosting my ego wasn't the main reason for taking up her time, I wanted to learn about the heart and specifically the electrocardiogram (ECG). 'We use ECG because it gives a snapshot of how the electrics of the heart are working,' explains Bastiaenen. She went on to tell me that the 'ECG is the most useful test we do, because it is so well established, giving us lots of data to draw upon'. ECGs can help in both long-term diagnoses and emergency situations. Her husband Nav (when not teaching handstands) is an emergency cardiologist: Bastiaenen tells me, 'ECG is one of the key diagnostic tools for a heart attack; it is the trigger that would get Nav out of bed at 2 a.m. to go to the hospital when he is on call.'

ECHO BASE

My experience of ECG began with me twiddling my thumbs in a hospital waiting room – never my favourite pastime, feeling hungry and slightly apprehensive about what it might show. I whiled away the time watching badly dubbed daytime television before being shown into a small room by an extremely charming ECG technician called Andrew who explained how it all worked. Reassuringly for my mental stereotype of hospitals, Andrew wore scrubs. The first thing I learnt about ECG is that it involves a lot of wires, two on the arms, two on the legs and six more that run from northwest to southeast across the heart (strictly the wires are called leads – this type of measurement is called a ten lead ECG). Having stripped off my shirt, Andrew attached electrodes via blue sticky pads. I then lay down, took a deep breath and it was all over. Waiting time one hour, ECG time five minutes – all pretty standard really. This underplays Andrew's expertise at putting the leads in their exact place, not an easy skill on people of all different shapes and sizes.

You can see my results in Figure 6 (next to a stylised version):

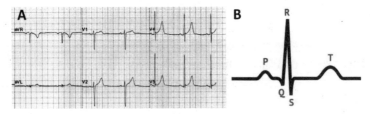

Figure 6. Echoes of former glory. A) My actual ECG plot (thanks to Andrew). B) An idealised plot showing the different parts of the wave that runs through my heart.

An ECG trace contains three components: the P wave (the initial bump), the QRS spike and the T wave – the subsequent hill. These waves relate to the way that electricity spreads through the heart. The spark starts in the pacemaker, located in the heart's top right-hand corner (hence the placement of the ECG leads on my chest). The pacemaker node works a bit like a hopper that slowly fills up and then empties all at once. This current is recorded as the P wave and then spreads through the tissue of the heart causing an ordered contraction of the different chambers. The QRS spike measures the contraction of the larger chambers (bigger on the chart because the chambers are bigger). The T part represents the heart returning to baseline ready to be shocked again. You will be reassured to know that my ECG had peaks in all the right places.

One of the biggest changes in recent years is that heart measurements like ECGs can be made in the comfort of our own homes and without needing an Andrew to place electrodes onto you. At the simplest level, Fitbits (and other similar wearables) measure heart rate. They do this through the little flashing light on their underside, using a technique called photoplethysmography. Blood vessels under your skin reflect the light, and the device interprets the changing strength of the reflections as the blood vessels open and shut. These devices can detect irregularities in heartbeats and alert the wearer that it may be time to speak to their GP. Devices that work by shining a light through skin do, however, have a potential ethnicity bias. The green light shining out of the back of the watch is more readily absorbed by the skin pigment melanin, which is higher in people with darker skin;[3] this affects the readings.

Cardiologists have access to far more sophisticated devices, allowing them to monitor over extended periods and detect changes day or night. One such device is the Holter monitor, a smaller unit worn around the neck for twenty-four hours. Note the spelling, not 'halter' (the horse collar). This is because the device was developed by Norman Holter, a physicist who worked on the atomic bomb tests on Bikini Atoll. The internet allows cardiologists to monitor the data remotely. This makes a significant improvement to quality of life. 'We can put patients in a virtual ward,'

Bastiaenen told me. 'This increases the time interval between hospital visits from a year to five years.' COVID lockdowns accelerated the rollout of this kind of approach. 'During the pandemic, patients who had had heart attacks could be prescribed medicine and monitored remotely to get the optimum dose without having to come into hospital.' Whilst micro trackers categorically were NOT injected in with vaccines, remote, online-monitored, personalised medicine is almost certainly the future of healthcare.

BUT HOW DOES IT WORK?

Having demonstrated that my heart pulses with electricity, it's probably worth a very quick lesson on the anatomy of the heart to explain what it actually does. In science writing, it can be difficult to gauge general scientific literacy and how to present information without being patronising. Luckily, my immediate family serve as a useful baseline: they are well educated, but have a history degree, a law degree, an engineering degree and a professional accountancy qualification between them (and to be honest a fairly sketchy understanding of science). Using the well-established research methodology of the 'family WhatsApp chat' I inferred that most people know that the heart comprises four chambers and goes 'boom boody-boom boody-boom'. The heart does indeed contain four chambers arranged in pairs (Figure 7): the right atrium and the right ventricle, the left atrium and the left ventricle. Figure 7 also shows how little it looks like the heart symbol, which came from either a plant, a French story about a pear or the mediaeval church, depending on which rabbit hole you descend on the internet.

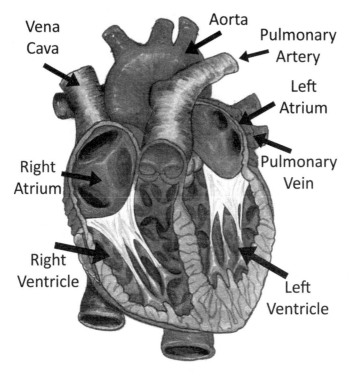

Figure 7. A heart. Four chambers of constantly pumping glory – not the seat of the soul though. Note, looks nothing like the heart symbol!

The atrium (literally the 'entrance hall') receives blood and then passes it on to the ventricle (which curiously comes from a Middle English word meaning 'little belly'). Ventricles then shunt the blood onwards. In humans the blood travels in two closed loops. To describe its journey, we will start at the right atrium (though as a closed loop you can start anywhere). The right atrium is fed by the vena cava, our largest vein, which drains blood from the body into the heart. As it travels around the body, the blood hands over its oxygen to the tissues, meaning it arrives at the heart deoxygenated. Once full, the right atrium empties its load into the right ventricle which in turn squirts it into the lungs via the pulmonary artery, thus completing the first loop. Having passed through the lungs it is now (hopefully) chock-full of oxygen. The blood returns to the heart at the left atrium via the pulmonary vein, gets pushed into the left ventricle and then fired around the body via the aorta. As I learnt getting my heart electricity measured, the atria and the ventricles contract at synchronised but different times (the boom-boom noise my sister and Peter Sellers described). One-way valves control the flow between the chambers, which allow the blood forwards but not back – a bit like a closed canal lock gate.

Not all animals have opted for the deluxe, four-chambered human model. Most insects (including the tiny fairy fly) possess just a single tube that pumps liquid around the body. Fish have a two-chambered system and frogs a three-chambered system. Octopuses pump their blood using three separate hearts. Curiously, horses require an additional four pumps to aid circulation. In humans, muscles help return blood from the peripheries. When you stand up or walk around, the contractions of your lower leg muscles push blood back up towards the heart (another reason that you should get up off your backside every so often). Surprisingly, horses lack muscles in their lower legs, and to compensate, they need more assistance to return the blood to their horsey hearts. This help comes in the form of a pump in each hoof (sometimes called the frog); weight put on their hooves squeezes the blood out of the frog and back up the leg.

The muscle type (cardiac) that surrounds the heart chambers differs from the muscle type (skeletal) found in your arm. This explains why squeezing a ball in your hand sixty times a minute is tiring, but your heart can do this right until the moment you die. The cells that form cardiac muscle – called cardiomyocytes – resist fatigue, partly because they contain more mitochondria.* Amazingly, at least half of the cells in your heart today are ones you were originally born with! The calculation for this is pretty cool. In the 1950s and 60s, the superpowers performed a series of nuclear weapons tests, including atmospheric bomb detonations. This led to a significant increase of the radioactive form of carbon (^{14}C) in atmospheric carbon dioxide (CO_2) – the normal form is ^{12}C; importantly, this increase happened only in a specific time window. Plants took up the radioactive $^{14}CO_2$, which through the circle of life eventually ended up in people's cells. The amount of ^{14}C in our DNA can be used as a time stamp; the heart cells laid down at birth in people born in the 1950s and 1960s contain higher levels of ^{14}C-containing DNA than ones that had grown later in their life. Using this method, scientists at the Karolinska Institute in Sweden determined the age of heart cells, demonstrating their remarkable longevity.[4] Unsurprisingly, cardiomyocytes (the muscle cells that make up the heart) eventually get tired. Cardiomyocyte senescence contributes to heart failure, where cardiac output fails to meet the body's requirements.

BROKEN-HEARTED

While most of the time our heart just ticks away without us paying it any mind, unfortunately, for a lot of people it does at some point go wrong. And when it does, it goes really wrong. The numbers for cardiovascular disease are HUGE.

* Did I mention that mitochondria are the powerhouse of the cell?

Globally, heart disease accounts for 17.9 million deaths a year, equivalent to the entire population of the Netherlands (or four times the population of Wales, to use the proper unit). Heart disease accounts for thirty-one percent of all global deaths: one-third of the people reading this book will die of something heart-related. Just in case these numbers aren't cutting through, another way of looking at it is in terms of frequency: around the world one person dies from cardiovascular disease every four minutes.

Cardiovascular disease doesn't just cost lives, it costs money, with astronomical yearly healthcare costs – every year in the UK alone heart disease costs £7.4 billion (or 24 billion Freddo chocolate bars), with £15.8 billion in associated costs. Which sounds a lot but can be hard to visualise. Luckily, the King's Fund provides us with helpful medical comparators: £1 billion pays for the running of one medium-sized British hospital for a year, or eleven thousand consultants or thirty thousand nurses.[5]

If you flick back to the introduction, you can see that, in the UK, heart conditions have been the main cause of death for everyone aged forty-five and over since the 1950s, replacing death from infection. However, in the last ten years, improvements in cardiology have reduced deaths from heart attacks. The endless zero-sum dance between the proximal causes of mortality continues, with deaths from cancer on the rise. There is no escaping death, and reductions in one terminal condition will lead to more deaths from another. The flip side is that conditions are interrelated; less cardiovascular disease means less stroke, less dementia and a better quality of life. Whilst one could argue that an acute, fatal heart attack after a life beats ten years of slow decline, that is a pretty high-risk strategy because many of the factors that contribute to heart disease act as repeat offenders in other diseases – drinking, smoking, poor diet, inactivity. All of which speaks to prevention being better than cure. Public Health England estimated that for every £1 spent on cardiovascular disease prevention, the return on investment is £2.30; that's a lot of red buses.[6] And sadly, there is inequality interwoven through health outcomes: the rate of heart disease is four times greater in deprived areas of the UK. I returned to Dr Rachel Bastiaenen to discuss problems of the heart. She told me about the three biggest cardiovascular problems: heart attacks, heart failure and arrythmia.

CORONARY IN THE COAL MINE

Heart attacks are caused by coronary heart disease, and they occur when the blood supply to the heart fails. Remember that the heart is basically a muscle that needs to pump all the time. To do all this pumping it needs oxygenated blood: some of the cardiac output goes directly back to supporting itself – it literally gets high on its own supply. Two large blood vessels feed the heart, encircling it like a crown, leading to their name – the coronary arteries, after the Latin. It is also why people sometimes refer to a heart attack as a 'coronary'; other names include 'myocardial infarction' for the medically minded. Simply put, a heart attack is when your heart suddenly fails. The heart stops because its blood supply stops – a really important distinction is that the vessels that feed the heart are blocked, not the arteries that exit the heart feeding the body. The main culprit for arterial blockade is atherosclerosis, caused by a build-up of fat on the artery walls. It is a bit like the enormous fatbergs they occasionally fish out of London's sewers

– the undisputed king of which was called Fatty McFatberg (obviously) measuring 250 metres long and weighing 130 tonnes. Fatty McFatberg took eight sewage workers several weeks to remove; they then converted it into ten thousand litres of biodiesel – a win, I guess? I'm not suggesting that people who have coronary heart disease have a 250-metre fatberg in their arteries, but on a smaller scale something similar happens, though with fewer baby wipes and condoms. The coronary arteries are approximately 0.5 cm in diameter,[7] about the thickness of a McDonald's drinking straw. The fatty plaques or atheromas reduce the diameter of these vessels, reducing the flow of blood and increasing the work needed to push it around.

One trigger for atherosclerosis is fat, particularly low-density lipoprotein (LDL) cholesterol in the blood. But just as London fatbergs contain a scaffold of other detritus around which the fat forms, atherosclerotic plaques aren't solely made of fat (Figure 8), they are scaffolded around immune cells.[8] These cells restructure the blood vessel, reducing the diameter. Inflammation from other sources (diet, smoking, drinking to excess) will fuel these immune cells to cause more damage. This process happens invisibly in the vessels feeding our hearts – until it crosses a sudden (and often lethal threshold). At this point, either the coronary arteries are so blocked that insufficient blood gets to the heart, or a plaque from elsewhere in the body detaches and plugs up the coronary artery. Both of these events are bad news. Basically, the heart, in the absence of oxygenated blood, stops working and dies. The signs of this include chest and arm pain. Pain in the arm may sound an odd sign of heart failure, but it relates to how our nerves connect the brain to the heart. This 'referred pain' occurs because our brains are unaccustomed to interpreting signals from the heart. Coronary heart disease is also experienced as shortness of breath, faintness and not infrequently death. About a quarter of first-time heart attacks are fatal.

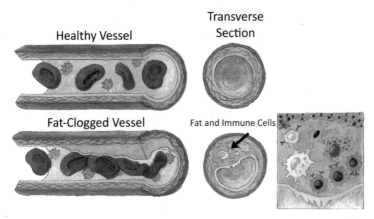

Figure 8. Fatberg in the heart. Blood vessels get filled up with fat, immune cells and other gubbins, which makes it hard for blood to flow. If this happens near a vital organ, the organ (e.g. the heart) stops.

TOTAL FAILURE OF THE HEART

Heart failure, the second major cause of cardiac complaints, sounds similar to heart attack but differs subtly. Heart attacks happen when the heart stops beating completely; heart failure occurs when the heart beats, but too weakly. Put simply, output does not match requirement. Heart failure can be caused by weakening of heart muscles due to ageing, genetic conditions or infection. Survivors of heart attacks often have some form of heart failure; if parts of the heart experience prolonged oxygen starvation, the muscles die and never grow back. Elevated blood pressure also contributes because the heart needs to work harder; like pumping a tyre, filling the last five percent of a full tyre is much harder than the first five percent when nothing is pushing back. In a similar fashion to heart attacks, ageing and inflammation both contribute to heart failure. As muscle cells age, they work less well and need replacing. In the context of chronic inflammation, connective tissue replaces dead muscle cells rather than more muscle cells. Ageing inflamed hearts become like cheap meat – shot through with fibrous tissue making them less supple. This stiffening of the heart necessitates more energy to contract and therefore reduces the power output. Whilst ageing is unavoidable, adding inflammation on top makes matters considerably worse. I asked whether one could reverse the damage, to which I got a long sigh from Bastiaenen. The short answer was no.

DANCING TO YOUR OWN BEAT

The third most common type of heart disease is atrial fibrillation, a form of arrythmia. Arrythmia relates to the electricity that flows through your heart and the need to co-ordinate the pumping of the different chambers. Because they act as a feeder tank to the ventricles, atria need to fill and then empty first. Occasionally the heartbeat can get all out of kilter. It can either race too fast (tachycardia, from the Greek *tachys* for 'rapid') or plod too slowly (bradycardia, from Greek *bradys*, which, you guessed it, means 'slow'). In atrial fibrillation, the rhythm in the upper chamber gets out of sync, most often going too fast. You may have experienced this as palpitations, which can be terrifying, especially when they happen for the first time. Dehydration can trigger arrythmia: I inadvertently tested this out in my twenties, by having a heavy night out, then a load of coffee to rebalance before biking to work. At the top of a hill my heart began racing uncontrollably. It did eventually calm down, but not until I had made a panicky call to Dr Rachel.* For most people this type of arrythmia is acute and self-resolving. However, for some people it continues. It isn't fatal in and of itself. But as Bastiaenen puts it: 'A heart that doesn't beat to a normal rhythm is never going to be as good.' The downstream damage depends a bit on which part of the heart is misfiring; ventricular fibrillation is associated with cardiac arrest, atrial fibrillation can lead to the formation of blood clots in the circulation, which are ticking time bombs.

FIX YOU

What can be done about heart disease? If you look around public spaces in the UK, you may notice the appearance of little boxes with a heart sign (which as we now know doesn't actually look like a real heart) surmounted by a bolt of lightning. You can use the website defibfinder.uk to find your nearest one, though I suspect that if you need one, this might be a bit late – and the website, quite rightly, directs you to call the emergency services. The boxes contain a device called a defibrillator, which you might recognise from your favourite hospital-based drama as the thing with pads that someone uses to jolt the patient back to life. Defibrillators completely reset the rhythm of the heart and kickstart it again. This can be lifesaving in some conditions – particularly ventricular arrythmias, where the big second

---

* Saving my life since 1995!

chamber, the ventricle, closes before filled fully, and thus fails to pump enough blood around the body. Scientists have been trying to reset the heart with electricity since the end of the nineteenth century; the first human was shocked back to life in 1947 (subtly different from resurrecting automatons made of a patchwork of dead criminals in German castles). Defibrillation doesn't work for all conditions. Fortunately, automated external defibrillators (or AEDs*) can read heartbeats by ECG before administering a reset burst. When used in the first three minutes of a heart attack, AEDs can increase the survival rate from fifty to seventy-five percent.

The type of condition that defibrillators work best on, ventricular arrythmia, is relatively rare. However, the man on the street can help during a heart attack with a much simpler intervention: CPR or cardiopulmonary resuscitation. This will also be familiar from popular culture – pumping the heart manually from the outside by pressing the chest; apparently the rhythm of the song 'Staying Alive' works best. You may have learnt CPR in a classroom using Resusci Annie. Incidentally, Resusci Annie is modelled on *L'Inconnue de la Seine*, an unknown, drowned French woman from the nineteenth century, who thanks to the CPR doll has probably the most kissed face in history.

CPR in real life is quite difficult, with a lot of force needed to pump the heart from outside. The chest needs to be compressed about five centimetres, which often leads to broken ribs. But as one internet guide stated: 'There are almost zero cases where a patient has significantly

---

* Not to be confused with IEDs; which are less helpful during a heart attack.

complained about CPR being performed on them,'[9] presumably because the alternative is 'or death'. A friend who performed CPR described it as extremely traumatic. The accident had crushed the victim's ribs, and my friend could physically feel the heart through the mangled chest, but ultimately they saved the person's life. If you find yourself in a similar position, do the same as them: phone the emergency services and stick your phone on speaker whilst following their instructions. The time interval is critical: the sooner you start the better and, unlike defibrillation, CPR works on all types of heart attack, as you basically replace the work of the heart. As Bastiaenen put it: 'If everyone knew how to do CPR, we would save more lives. Most cardiac arrests happen outside the hospital.' Her advice is: 'Think Vinny Jones.'*

Stopping people dying on the street is a good start, but what happens when they get to hospital? Fixing broken hearts is one of the success stories of twentieth century medicine. The incidence of cardiovascular disease peaked in 1968 and has been declining since then.[10] A lot of this is due to prevention, but amazing advances in surgery should also give us heart.

The first human heart surgery (depending on how you define it) was performed sometime in the early 1800s, when two doctors, Romero and Larrey, operated on the sack around the heart. An African American surgeon called Daniel Hale Williams performed one of the very first surgical operations directly on the heart itself. Williams, born in Pennsylvania in 1856, worked as an apprentice shoemaker, then a barber

---

* Though I am not entirely clear how grabbing the person's nuts helps.

before setting up a medical practice in 1883. In 1893, he successfully performed reparative surgery on one James Cornish, stabbed in the heart during 'an altercation'.[11]

Since then, heart surgery has leapt forward in incremental bounds. However, the heart posed a bit of a problem to cardiology pioneers because of its fundamental role in keeping us alive. Operating on other organs is (somewhat) more straightforward: you can for example isolate someone's liver for a couple of hours without them noticing; however, we can't survive very long at all without our hearts. The received wisdom is that less than four minutes without a heart is recoverable, longer than ten minutes is fatal. An Italian mountain climber called Roberto holds the unenviable record of the longest time spent in full cardiac arrest at a remarkable eight hours forty-two minutes. He was caught in a freezing waterfall during a thunderstorm whilst climbing Marmolada, the highest mountain in the Dolomites. This triggered hypothermic cardiac arrest – basically a frozen heart. But rather than let it go, he was rescued, flown to Treviso hospital and once warmed up given a dose of electricity from a defibrillator, recovering completely with just a mild dose of amnesia.

Since freezing patients in waterfalls is not entirely practical, to operate on the heart surgeons needed to find a way to provide oxygenated blood whilst the heart of the person under surgery remained out of action. The first attempts proved unmitigated disasters: of the eighteen trials reported between 1951 and 1955, seventeen led to death. However, surgeons persisted. Clarence Walton Lillehei tried the approach that worked for 'lucky Roberto' – inducing hypothermia to reduce the patient's need for oxygen – but this gave only a short window of about ten minutes, not long enough for intricate surgery. He then tried co-circulation, plumbing the blood vessels of one healthy person into another for the duration of the surgery, which failed in twelve out of forty-four operations. Co-circulation did not lack risk – why accidentally kill one person in surgery when you can kill two?

The American surgeon John Gibbon provided the solution. During the 1930s Gibbon witnessed a failed pulmonary embolectomy (removing a clot from the lungs), which inspired him to develop a machine that could substitute for the heart. In partnership with his wife Mary, Gibbon developed a prototype heart-lung machine. Gibbon's research underwent a brief hiatus whilst he served as a Second World War surgeon in Burma (now Myanmar), but once the war ended, Gibbon received funding from IBM to develop his machine further. By 1953, Gibbon operated on a patient with a hole in their heart. He succeeded, keeping them alive for twenty-six minutes of full bypass whilst he performed the reparative surgery.

One of the repair operations that Gibbon's heart-lung machines facilitate is slightly confusingly called a heart bypass. It is not actually the heart being bypassed but the arteries that feed it. As seen, in cardiac artery disease (the most serious form of heart attack*) the arteries that keep the heart alive can get clogged up. To fix this problem, surgeons take an artery from another part of the body and sew it back into the heart, thus restoring blood flow; often they need to transfer more than one artery. Four is fairly normal; there is at least one instance of someone having

---

* 'Most' being a bit subjective here; if your heart stops, things aren't going well whatever the reason.

fourteen reconnected,[12] which a surgeon friend referred to as being 'a bit excessive' since the heart is usually fed by only ten major arteries. One challenge is finding enough working arteries in the patient to replace the damaged ones. However, we find ourselves entering an age where new vessels can be grown in a lab from the patient's own cells before the operation, though this is some way off being universally available.

It is now possible to replace or repair nearly every part of the heart, including the sinoatrial node. Whilst defibrillators can reset acute arrythmia, a longer-term fix can be required. This is where the artificial pacemaker comes in. When not connecting two people together, the heart surgeon Walton Lillehei also developed an externally worn pacemaker. This was a box the size of a Walkman* that you could wear on your belt, with leads that ran through your chest wall directly into your heart. Whilst effective, it was clearly a bit impractical. Around the same time, a Swedish duo, Elmqvist and Senning, developed an implantable pacemaker. Their first device failed in three hours, the second one (implanted in the same patient) lasted two days, demonstrating proof of principle. Remarkably, the first recipient of an internal pacemaker – Arne Larsson – went on to have a further twenty-six different pacemakers fitted and outlived both surgeons. The modern iteration is about the size of a car key and sewn into the heart. Over one

---

* For those of you not children of the eighties, the Walkman was a portable cassette tape player – the forerunner of the iPod, which in turn was the forerunner of the iPhone (ain't getting old fun?). Presumably if this goes to a second edition, I will need to explain what an iPhone was. For reference, a Walkman was about half the size of a paperback book (a book is the forerunner of the Kindle).

million pacemakers are fitted a year leading to a significant improvement in quality of life.

Sometimes, however, the heart is just too knackered to carry on and no amount of replacing valves, arteries or fitting new batteries will help. The whole thing needs replacing, by transplantation. The idea had been around since 1945 when a Soviet scientist, Nikolai Sinitsyn, transplanted a heart from one frog to another frog and then from a dog to another dog (I'm not sure why he stopped there and didn't move onto a hog; Dr Seuss certainly would have and he didn't even finish his PhD). Christiaan Barnard performed the first human-to-human heart transplant in South Africa in 1967. Whilst much of the pioneering research was undertaken in the US, getting hold of a living heart to transfer proved problematic. Taking the organ from one person for another person to live is complicated ethically, and defining death is tricky. Laws in the US around death in the 1960s related to cessation of the heart, making it impossible to perform transplants (because the heart was dead). In South Africa death was defined by brain death, which meant the heart could remain alive. The first heart transplant only lasted for eighteen days, with the patient Louis Washkansky dying of pneumonia. The main problem for transplant was rejection by the immune system; we will return to this when thinking about kidneys, but suffice it to say, this can now be overcome.

FIX ME

My hope, however, is to not need any of these interventions. How can I avoid heart disease as I age? There isn't much I can do about my genetics, yet. Instead, I need to focus on environmental exposures and risks. I can helpfully tick off one right away as I don't smoke. I feel healthier already.

The heart presents me with the easiest intervention to delay the impact of ageing. Do more exercise. And it is such an easy win (or should be). If you think about the heart purely as a mechanical device then doing exercise to make it better can seem counter-intuitive. This is where analogies with mechanical, human-built machinery fail. The more you run a car engine, the less well it works over time, but biological tissues are adaptive: the more they work, the fitter they get. Since the heart is a muscle, working it (in short bursts) improves function rather than degrading it. Bursts of exercise improve the quality of the heart, increasing the number of the cardiomyocytes, keeping it flexible and also increasing the volume of blood pumped per beat. The increased volume reduces the need to beat as often; consequently, it works less hard at rest. Exercise provides a slew of other benefits for the body that help the heart indirectly: it reduces weight, thereby reducing load on the heart; it modifies inflammation (our go-to villain); it improves the efficiency of the muscles to get oxygen – reducing the overall workload on the heart; it returns blood from the extremities; it increases cellular repair; it burns glucose; and it reduces cortisol (the stress hormone). Because everything is interrelated, exercise reduces diabetes, stroke, cancer and even mental ill health.

When it comes to improving life outcomes, exercise considerably trumps nearly everything I am planning to do whilst writing this book. And with this in mind, when I started to write this chapter, I had great visions of turning over a new leaf exercise-wise. But then I pulled a muscle in my calf, and being a man of a certain mindset didn't let it recover and pulled it again, and then again.* I had visions of sharing my recorded mileage, and the transformational nature of running apps like Strava. The sad truth is that, because of my injury, in the month when I wrote about the heart I cycled for a total forty-five minutes, I ran for thirty minutes (the grand distance of four kilometres) and I walked for four hours; I also did some gardening at the allotment and low-level rowing on a machine. Altogether this felt like I had failed. However, adding it all up, it worked out as about 150 minutes of moderate exercise per week; the NHS recommended minimum limit. This feels like the exercise equivalent of 'five a day'† – a relatively low bar, better than nothing but maybe less than ideal. The 'recommended' ten thousand steps a day is also somewhat arbitrary; it derives from the Japanese symbol for ten thousand, 万, which looks like a man walking. Recent research demonstrated that the exercise bar really is this low, just three bursts of high-intensity activity a day can reduce the risk of all-cause mortality.[13] This type of vigorous intermittent lifestyle physical activity (VILPA) can be as little as running after the bus or walking up the stairs. In case you wondered, sex probably only counts as moderate exercise and masturbation as light exercise (unless you undertake either very vigorously).

We really don't need to do much exercise to see improvements in outcome; if I remember to walk up to my office on the fourth floor more than once a day then I am in the gravy (or out of it I guess). One of the (many) downsides of the

---

* Ironically, I then pulled the same muscle yet again whilst doing structural edits on this chapter a year later.
† The suggested minimum daily intake of fruit and vegetables, also surprisingly low.

COVID pandemic was that working from home made everyone (even) more sedentary. A normal working day (pre-pandemic) at least comprised walking to meetings, chasing buses, going to the tearoom; in the endless Zoom nightmare, people basically became rooted to one spot. Sitting still is not great. A landmark study in the 1950s found that double-decker bus drivers were twice as likely to suffer heart attacks as the conductors – because the conductors climbed six hundred stairs a day, up and down the Clapham Omnibus. Working in a lab is cardioprotective (for me at least), because I spend a large amount of my day walking around the lab trying to find my team who possess a remarkable talent for hiding from me; I tell myself it is because they are busy.

Moderate exercise is defined by increased breathing, but still being able to talk, whilst vigorous means fast breathing and difficulty talking. In heart rate terms, moderate is about seventy percent of maximum heart rate and vigorous about eighty percent. Heart rate maxes out at 220 bpm minus your age; for someone in their mid-forties like me it goes max 180, vigorous 145, moderate 125. Returning to my star jumps at the beginning of the chapter, they did successfully raise my heart rate, but not enough to count as moderate exercise – the PTI would be furious (bless his tiny little tank top and quivering moustache).

Doing too little exercise causes multiple problems, but what about the other end of the spectrum? In terms of time efficiency, why not push the heart even harder? One change in the last thirty years is the age into which people carry on doing high-intensity exercise and what that means in terms of health. Whilst not quite matching the extraordinary feats of Jimmy Anderson, still (at the time of writing) wheeling in at forty-two, twenty-one years and seven hundred wickets after his Test cricket debut, my generation (Gen X) continues to run marathons and don Lycra to block up the roads in pelotons way after my parents had opted for a bit of gentle gardening and walking groups. The impact of this on overall health remains to be seen.

But if you are not regularly doing high-intensity exercise, ease yourself into it, as over-exercising takes longer to recover from; one mega session a week provides less net benefit than several lower intensity ones. I found this out in my mission to get fit for writing. By overdoing the exercise one time, I became unable to exercise properly for a month. I guess I have inadvertently proven a point more effectively than just telling you I smashed a few personal bests and could still run faster than my friend Alan and his skinny ankles. Exercise is great, but sadly, as we get older, we can no longer spring straight from the couch to winning the marathon. This relates to inflammation and a reduced capacity of our cells to repair themselves. My torn/damaged/slightly hurty calf muscle is not getting better as fast as I remember because I am getting older. One other thing I can't avoid is a general downward decline in cardiac output: the muscle cells in my heart are, after all, as old as me and beginning to slow down.

But inevitable decline aside, with minimal input I have sorted my heart health and persuaded you to put the book down and do some physical jerks. We have covered the importance of the heart, how it goes wrong and what can be done to prevent this, or if not fix it. We've also seen how, of the big four, exercise contributes to heart health. Through a combination of prevention, drugs and surgery, heart disease is very much on the decline – so much so that I have struck it off the top of my list of things to worry about. One area I've not really touched on is *what* the heart actually pumps. And once you return from your moderate-intensity workout, in the next chapter we will look at blood and circulation. And, critically from my perspective, why high blood pressure causes so many problems.

CHAPTER 5

A Stroke of Bad Luck: Killer Number 2 – Strokes

'Blood is a very special juice.'

Goethe, *Faust, Part One*

Second place in the general list of killers, but topping my own personal harbingers of doom, is stroke, caused by high blood pressure. Stroke and heart disease are intertwined, because of the central role of the heart's output, the blood.

Now that we turn to thinking about blood, I cannot ask you to self-experiment. Even as a working research scientist, I'm forbidden from experimenting on my blood. The risk comes from altering the blood and then accidentally injecting it back into myself; if I needle-stick with someone else's blood, my immune system will recognise and reject it, but my own (now modified) blood will be welcomed with open arms by my body, harbouring whatever nasty thing I'd put into it. That doesn't stop other researchers asking us for our blood – in fact, the restriction on doing our own research means that donating blood forms part of the social contract of a medical lab. We literally bleed for the job. It's not all bad: I met my wife giving blood for one of her studies. But it can occasionally go awry. One time, a colleague somehow forgot to connect the vacutainer onto the tubing that ran from the needle in my vein, thereby creating a sprinkler effect with my blood and redecorating the lab with serial killer vibes. Unsurprisingly, that put me off donating for some time. For reasons that remain unclear to me, another colleague wanted to see if they could measure antibodies in the fluid lining the urethra. The method of collecting the sample involved passing a small piece of blotting paper down the hole at the end of my penis; genuinely do not try this at home. Science certainly has its ups and downs.

The blood vessels serve as the motorways, trunk roads and tiny country lanes of our body. Blood itself makes up the traffic that passes down these fleshy highways. Once our ancient ancestors evolved from the single-cell stage, they needed a way to move stuff around their proto-bodies. All multicellular organisms use something like blood: trees swell with sap and insects have haemolymph (pumped around by their teeny tiny hearts). Blood's main cargo is oxygen, but it also delivers sugars and nutrients from the guts to everywhere else and removes waste products via the liver and kidneys. Our blood vessels also carry the body's emergency services (immune cells) to sites of infection. Finally, blood contains an inbuilt motorway repair system, clotting, which plugs any leaks; however, clotting can also cause major problems by disrupting flow – not dissimilar to roadworks.

To do all this, an average-sized adult contains about five litres of blood.* This amount varies with height, weight and sex (you can roughly estimate your own blood volume using the calculation: seventy-five millilitres of blood per kilogram of body weight for men and sixty-five millilitres per kilogram for women; for more accuracy you need special

---

* Ten pints in these post-Brexit days, probably with a crown on.

heavy water and a mass-spectrometer). We can lose fourteen percent of our blood without really noticing,[1] allowing you to donate 470 ml (about a pint) at any one time. This would be categorised as a mild haemorrhage ('haemorrhage' from the Greek words for 'blood' and 'burst'). Losing more than fifteen percent begins to get a bit serious, leading to nausea, fatigue and a cooling of the extremities as your body redirects blood to the parts that need it most. More than forty percent blood loss and things get extremely severe; the body enters shock because the blood fails to meet the oxygen demands of the body. The good news is that your body can replace lost blood. Your body makes two million red blood cells a second; since blood contains four million red cells per millilitre, in theory it can replace a pint in fifteen minutes. But red blood cells also constantly turn over, so it actually takes about one month to fully replace a pint of blood.

Why do we need trillions of red blood cells? Without them we would die from oxygen starvation. Erythrocytes (from the Greek *erythros*, 'red') shuttle oxygen around the body; and as everyone should remember from the depths of their science memory, they are shaped as biconcave discs, the only other recorded object with that shape being the Refresher sweet from the 1980s (the sugary disc ones, not the chewy strips). Red cells contain haemoglobin which picks up oxygen in the lungs and deposits it where needed. To improve their oxygen carriage efficiency, during evolution red blood cells lost ninety-nine percent of the things that we associate with human cells; they don't have the cellular powerhouse (the mitochondria), ribosomes or a nucleus – only haemoglobin. This means that curiously red blood cells don't use oxygen themselves.

Red blood cells are just one part of the blood; it contains three other components: plasma (the liquid bit), white cells and platelets. Plasma makes up most of our blood, about sixty percent of the total volume. It is about ninety-one percent water; the rest is salt and protein. The proteins belong to three main categories: plasma proteins, clotting agents and antibodies. Blood also carries smaller amounts of other proteins that signal to the rest of the body, for example the cortisol and adrenaline which affect the heart rate. The eponymous plasma proteins help maintain the blood at the right pH (level of acidity) and concentration, the technical term for which is 'osmolarity'.

The white blood cells are the immune cells. In my unbiased view as an immunologist, they are clearly the most important and most numerous cell in the body[*]. There are somewhere between 1.8 trillion[2] and 3.4 trillion[3] white blood cells in an average man, but they only contribute 1.2 kg to centrist dad's 72 kg body mass; most of the mass is muscle cell (even allowing for middle-aged spread). If you were also an immunologist, I would now be arguing with you about the best immune cell (and why it is the T cell).

THE CLOT THICKENS

Last but not least are platelets, also called thrombocytes from the Greek *thrombus* meaning 'clot' and the ubiquitous *cyte* meaning cell. Giulio Bizzozero first identified platelets as an independent blood cell in 1882.[4] He described them in

---

[*] After red blood cells, which don't really count because of their lack of anything other than haemoglobin.

a scientific paper first in Italian, then French and finally German – an impressive achievement. I find it hard enough writing scientific papers in my native English. He named them *piastrine*, *plaquettes* and *blutplättchen* in the three corresponding languages. Also referred to as cellular dust, platelets are cell fragments, about one third the size of an erythrocyte. The human body makes around 10^{11} platelets every day (about the same inconceivably large number as red blood cells).[5]

Platelets repair damaged blood vessels, working with the clotting factors found in plasma. Repairing damaged blood vessels is vital, not just for surface cuts, but internal bleeds (that lead to bruises). In most of us, this happens without a thought, though in big bleeds we might need to assist with pressure or stitches. However, for 0.05% of the population small cuts can cause fatal consequences. Haemophilia stands out as the best-known condition affecting the clotting process. People with haemophilia have a mutation in one of the (many) clotting factors found in the plasma. The most famous example of haemophilia is in European royal families in the nineteenth and twentieth centuries. Queen Victoria appears to have developed a spontaneous mutation in clotting factor IX. She passed this on to two of her five daughters, Alice and Beatrice, who didn't manifest the condition, and her son Leopold who did. Clotting factor IX is on the X chromosome; princes have only one copy of this, princesses two, hence the male progeny were affected, whilst the females only carried it. The tendency of European royals to marry their cousins ensured the spread of Vicky's dodgy genes into the Spanish, German and Russian dynasties. Some people argue it contributed to the fall of the Romanovs.

The process of forming a clot is not trivial. Biomedical textbooks describe it as a cascade – another way of saying look away now. I cannot begin to come up with a simple way to describe it. Twelve clotting factors (found in the plasma, all with roman numeral names, but not in numerical sequence) interact in a series of events that makes the procession of kings in the Wars of the Roses look simple.[*] The net result of clotting is that an inactive protein called fibrinogen gets cleaved into the active fibrin form. Fibrin is sticky, fibrinogen isn't. The sticky fibrin protein glues platelets together over the damaged part of the blood vessel, forming a net that traps other platelets, red blood cells, etc. In our fatberg analogy, fibrin acts as the baby wipes flushed down the sewers. A clot eventually leads to a scab, which in turn leads to the best thing in the world, picking off a scab (but only when it is ready and never before!).

Why, might you ask, have we evolved such a hideously complex clotting system – apart from to mess with medical students in their first-year exams? It is because misplaced clots are very bad for us. The problem isn't that the clots form in the first place, but where they end up. They can become dislodged and then float around the body until they block something more important. Deep vein thrombosis (DVT), where a clot forms in a remote vein and sloughs off, causing trouble elsewhere, illustrates the dangers of misplaced clots. DVT rose to public awareness through an association with long-haul flights and the

---

[*] Since you (never) asked it goes: factor XII is activated to XII_a, that cleaves factor XI into Xi_a that cleaves factor IX into IX_a that cleaves factor X into factor X_a that cleaves prothrombin. P. S. the kings go Henry, Henry, Henry, Edward, Henry (the same as the one before Edward), Edward again, new Edward, Richard, New Henry.

suggestion that wearing tights might in some way help. But it turns out flights are an 'often cited, yet relatively uncommon' factor,[6] occurring in fewer than five passengers for every million air travellers flying over ten thousand kilometres (slightly further than London to Cape Town)[*]; so unless you find yourself heading to the *Rocky Horror Picture Show* at the Sydney Opera House, you can remove the stockings and suspenders. The more serious common factors for DVT should be getting familiar by now: age, immobility, smoking, obesity. Surgery and birth control pills can also lead to unwanted clots, as can pregnancy and birth.

How do clots get from where they form to where they cause damage?

CIRCULATION

Blood travels around our body in a network made of three major types of vessel: arteries, veins and capillaries (Figure 9). Arteries travel away from the heart; veins return blood to it. The heart provides the motive force to move blood around the body. Firstly, blood passes through the lungs, dumping carbon dioxide and picking up oxygen, which it then delivers to the organs that need it. Muscles surrounding the veins squeeze oxygen-depleted blood back towards the heart, and valves within the veins prevent blood from going back in the wrong direction. If these valves fail, blood builds up in the lower body and causes the leg veins to swell up – varicose veins. The treatment for which is basically to burn and seal defective veins.

---

[*] Or fifty times the length of Wales using the proper units.

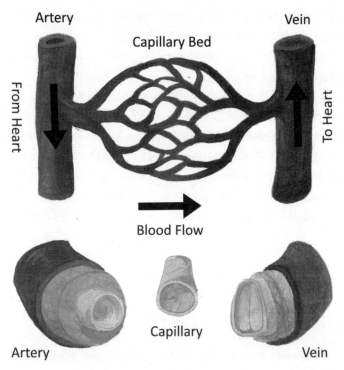

Figure 9. Vessels large and small. There are three types of blood vessel: arteries, capillaries and veins. Top image shows the connections between the vessels; bottom image cross-sections. Cross-section shows how capillaries are much thinner-walled.

Capillaries are the tiny vessels that connect arteries and veins. Gaseous exchange in tissues takes place via the capillaries. To facilitate this, they are extremely thin and are lined with subtly different types of cells: epithelial cells line arteries and veins, endothelial cells line capillaries. Simplistically, epithelia stack vertically, while endothelia connect horizontally, creating a thin cell layer that makes it easier for gas particles to reach tissue. Capillaries can easily become damaged – spider capillaries or telangiectasia is the condition seen on the noses of Tory grandees and anti-quated members of the MCC. A range of reasons cause these, but alcohol and a general anger about the short format game don't help.

Because the capillaries are small, thin-walled and under pressure, liquid gets forced out of them and into surrounding tissues. This is beneficial because the things dissolved in plasma (sugars, etc) can get to the tissues, but it involves large volumes. About twenty litres' worth of fluid are pushed out of the capillaries each day (bearing in mind we have only five litres of blood in our body), so we need a way to recover this. About three-quarters of the extruded liquid drains back into the capillaries, but the remaining quarter needs to be recovered through a different system, the lymphatics. This acts like a ghost fluid recovery system, returning plasma to the blood circulation via two ducts that feed into the large vein running directly into the heart. Curiously, the ducts don't evenly split the workload, the right duct drains half the head and the right-hand side of the torso, the left does everything else including both legs. It also incorporates a surveillance system; the lymph nodes that swell up during infection are training your immune response.

Whilst the plumbing diagram of the human body is now pretty well understood, it took a surprisingly long time to be established. The idea that blood flows in a closed circuit is often attributed to William Harvey, who served as James I's private physician, but the Arabic scholar Ibn al-Nafis published a work that pre-dated Harvey's by four hundred years. Born in Damascus in 1213, Ibn al-Nafis lived most of his life in Egypt. He wrote a manuscript entitled *Sharh tashrih al-qanun li' Ibn Sina*, or 'Commentary on Anatomy in Avicenna's Canon', only rediscovered in 1924 by an Egyptian scholar. In his commentary, al-Nafis proposed that the blood did indeed flow from the heart into the lungs and back again. Whether this work influenced Harvey is unclear, as is the question of why his contribution remained unacknowledged for seven hundred years, or why the manuscript turned up in the Prussian State Library, Berlin – the Iron Chancellor not being renowned for his health tips.

ROGUE CLOTS

The circulation of our blood is clearly beneficial, but it also enables the movement of clots, often with catastrophic results when the clot stops blood flow to a vital organ. Misplaced clots cause a quarter of all deaths. In addition to contributing to coronary heart disease, unwanted clots cause two other major conditions, pulmonary embolism (blocking blood to the lungs) and strokes. Of the two, strokes are more common.

Strokes occur following disruption of the blood flow to the brain. There are two types, ischaemic and haemorrhagic, derived from Greek words meaning 'restraining' (*iskho*) and 'blood' (*haem*). Clot-associated strokes are the ischaemic ones; haemorrhagic strokes occur when vessels that feed the brain burst and blood that should be supplying these areas leaks out. Strokes lead to a huge number of deaths – nearly as many as heart disease. Globally it is the second-leading cause of death. The CDC estimates that every forty seconds someone in the US has a stroke and that someone dies from them every three and a half minutes, costing more than $50 billion.[7] And of course, death isn't the only outcome: strokes can cause severe long-term brain damage, with the type of lost function depending on which part of the brain becomes oxygen-starved.

Disrupting the blood flow to the brain causes rapid and incalculable damage because our brain cells are incredibly hungry; the brain burns twenty percent of our total energy. In the absence of oxygen, brain cells keep trying to burn sugar but start producing lactic acid (the same compound that gives your muscles a burning feeling when sprinting). This causes further cellular damage. As can be imagined, the sudden and widespread death of brain cells is not a good thing. Treatment very much depends on the type of stroke. Ischaemic strokes need the clot to be broken up to restore blood flow; haemorrhagic ones need leaks to be fixed. As with heart injury, speed is critical: the longer a tissue goes without oxygen, the less reversible the damage.

The rate of stroke dramatically increases with age. Risk factors naturally include obesity, smoking and alcohol, but stroke also has some additional, subtly different risk factors, particularly sex-specific ones – both pregnancy and taking oral contraceptives increase the risk of clotting and strokes (as with DVT). Oral contraceptives contain progesterone and oestrogen. High-dose oestrogen has an association with an increased risk of thrombosis (clotting) because oestrogen increases the levels of clotting factors – possibly as an adaptation to menstrual bleeding. But it should be noted that the absolute risk of clotting from oral contraceptives is low (60 in every 100,000 women), and the danger is much less than the pregnancies they prevent, which can increase the risk of strokes by up to fivefold.[8] Inflammation is another contributory factor, doing double damage; it increases the risk of strokes happening and then exacerbates them when they do. Chronic inflammation reduces the diameter of our blood vessels, increasing the likelihood of clots sticking. And when blood vessels pop, an already inflamed immune system will amplify damage done to brain cells, leading to a cascade of destruction.

The two biggest risk factors for strokes, however, are high blood pressure (hypertension) and high blood cholesterol (hypercholesterolemia). This was a bit concerning for me, as my health check reported that both values hovered higher than they should.

UNDER PRESSURE

The oxygen needs of our organs are not uniform: when running, your legs need more blood; when digesting a meal, more goes to your guts. Our bodies have evolved ways to shunt blood to places it's required. Blood vessels contract and expand, redirecting the flow. They also react to changes in temperature: closing down when cold, allowing us to conserve heat; opening up when hot, allowing us to shed heat. The pink face from overheating and the flush of embarrassment have similar mechanisms – vasodilation of capillaries in the face. People of all skin colours blush, though melanin can mask it.

To visualise the effects of vessel contraction on blood pressure, let's return to the experiments. Put a drinking straw in a liquid and blow bubbles through it; now repeat but pinch the tube. The effort to make the bubbles increases as the vessel thins. Vessel contraction leads to redirection because the blood will take the easiest route. There is a differential in pressure between the arteries and veins – the blood leaves the heart at a higher pressure than the blood returning, the energy having dissipated throughout the whole body. Most of the blood in our bodies resides in our veins, it squirts out of the heart fast and then trickles back slow. A red blood cell takes about a minute to complete a circuit of the body. The blood pressure we record, strictly the arterial blood pressure, reflects all of these changes.

How do we measure blood pressure? At some point you will (like me at the testing centre) have probably had your arm put in a cuff attached to a machine that squeezes it uncomfortably and then releases it. This is catchily called a sphygmomanometer, a horrible portmanteau word combining the Greek for 'pulse' with the French for 'pressure meter'. It captures two values called the systolic and diastolic blood pressures. The systolic (higher number) represents the force at which the heart pushes out blood; the diastolic value reflects the pressure of the system at rest. These measurements are determined by sound in the blood vessel. The cuff increases the pressure in the arm until heart noises disappear; following the cuffs' slow release the systolic value is the level at which heart sounds return and the diastolic is the pressure at which they fade away again.[*] The detected lub-dub noises are called Korotkoff sounds after Nikolai Sergeyevich Korotkov, the Russian surgeon who first described them. Korotkov survived the Russo-Japanese War, the First World War and the October Revolution, only to be carried off by TB in 1920. In the days

---

[*] The reason why sounds appear and disappear relates to the movement of blood in the veins. Normally, blood streams soundlessly through our veins; if partially constricted, flow noise increases due to turbulence. At the top (diastolic) pressure, flow stops completely (no noise); releasing the pressure increases the noisy flow. At the bottom (systolic) pressure, normal, silent service is restored. Thanks to Dr Vivek Muthu for explaining that over an excellent meal.

before automated machines, Korotkov's sounds were determined by stethoscope – the status symbol of medical students everywhere. Blood pressure units are given as millimetres of mercury (mm Hg) that could be lifted by that amount of force.

Recorded systolic and diastolic numbers can then be compared to a standard set of values based on huge amounts of accumulated data that tell naughty blood pressure from nice. Goldilocks-like, there is a sweet spot, with a U-shaped distribution of increased risk of mortality from the optimum centre point value – being too high (hypertension) or too low (hypotension) increases the likelihood of death. Given the bad press hypertension gets, it was surprising that its sneaky sibling hypotension is just as likely to cause you harm. If the higher (systolic) number is about 120 and the lower (diastolic) number about 80 then everything is A-OK. For hypertension, anything above 140 is bad and above 180 is extremely bad. Hypotension begins when systolic pressure falls below one hundred.

Hypertension puts the whole circulatory system under stress. Your blood vessels can cope with increased load for short periods of time, but under sustained pressure they may burst, which can cause a haemorrhagic stroke. Meanwhile, repairing blood vessels can lead to clots, which can lead to an ischaemic stroke if they float off and get lodged somewhere important. Damage to the blood vessels feeding the brain leads to haemorrhagic strokes. At the other end of the scale, hypotension is dangerous because the blood may not be delivering enough oxygen to the places needed, leading to dizziness, fainting and falls.

SALT, FAT, STROKE, DEATH: MASTERING THE ELEMENTS OF HIGH BLOOD PRESSURE

In order to know how to fix high blood pressure, it helps to understand how it can become elevated. There are some underpinning genetic conditions, but as with much of the conditions that deteriorate as you get older, most risk factors are lifestyle-associated. Collectively anything that alters the amount of blood or the pipes through which it is pushed alters blood pressure. Obesity, for example, causes layering of fat within the arteries, decreasing their aperture and increasing the force needed to pump the same amount of liquid through them. Smoking raises blood pressure through one of its (many) unpleasant ingredients, nicotine. Dehydration and the consumption of things that dehydrate you (caffeine/alcohol) increases vasoconstriction, increasing pressure.

Importantly, dehydration increases the concentration of salt. Salt, as you may be aware, is a key player in the blood pressure game, especially when overconsumed. Salt, or sodium chloride (NaCl) to give its full name, needs tight regulation in the blood. Blood should be about as salty as sea water. NaCl accounts for 0.4% of our total body weight; an average white adult man* contains approximately three hundred grams or about 130 teaspoons of salt. As its name suggests, NaCl is made up of two elements, sodium and chlorine (in their ionic, or charged, forms – Na^+ and Cl^-).

---

* This is a good example of where, by focusing only on white men, research fails other groups. It is not known whether women are more or less susceptible to the effects of salt and how this changes around menopause. Likewise, racial differences may alter the impact of salt.

Whilst cardiologists debate the exact mechanism by which elevated salt causes disease, sodium should be considered blood pressure enemy number one. The most widely accepted mechanism is that increased sodium increases water retention, increasing the amount of liquid that needs pumping around the body and therefore the pressure. Under normal circumstances, increased sodium in the blood indicates to your body that you are dehydrated. The body reacts by releasing a substance called antidiuretic hormone (or vasopressin) which we will return to when we think about the kidneys. However, if you have just eaten a bag of crisps, you confuse the sensors in your body; hence why pubs serve salty peanuts – they increase thirst. Sodium also reduces the ability of the sympathetic nervous system (the fight or flight one) to relax, leaving everything more tense, explaining the junk food jitters. Either way, salt intake and blood pressure are directly related, and elevated blood pressure correlates with all kinds of poor health outcomes.[9]

Since hypertension is so bad, how do we lower it? The simplest way is to reduce salt intake. This doesn't even have to be done for a long time to see effects – within four weeks of salt reduction blood pressure drops.[10] There are (inevitably) variations in how much salt is acceptable, but the average amount is one teaspoon a day (2.5 g); there may be a bit of leeway, and a recent study has suggested bad effects kick in at around the two teaspoons (5 g) a day mark.[11] But before you go crazy and sprinkle the Saxa all over your chips, know that much of the food we eat, especially junk food, already contains perilously high levels of salt – a single pizza can exceed the daily limit. However (for me), there is a huge gulf between the good intention of salt reduction and the practice. After various rounds of nagging and a few run-ins with the nurse, I tried to reduce my salt intake. The net result: duller food and no noticeable change in my blood pressure.

I constantly skirt the boundary between healthy and unhealthy blood pressure, particularly my upper (systolic) number which on a good day sits just south of the 130 mm Hg threshold. However, due to pride (and denial) I am not yet ready to embrace middle-aged pill-popping – so I am extremely receptive to alternative therapies when it comes to hypertension. One suggestion is to listen to twenty-five minutes of music a day – though the study only tested classical music;[12] the impact of the *Moana* soundtrack, Taylor Swift and nineties-era dad rock have yet to be recorded.

Another is to eat more bananas.* Bananas are rich in another metal ion, potassium. When you eat a banana, the potassium dislodges sodium from the blood. However, the high levels of potassium in bananas makes them very slightly radioactive, because one of the naturally occurring forms (isotopes) of potassium (^{40}K) is unstable. A truck full of bananas contains enough radiation to exceed the annual safe UK dose (2,700 μSv), and it would take a literal boatload to kill you. So, don't fear the banana: the amount of potassium in an average person (140 g) far outstrips that in a banana.

One food has the completely opposite effect to bananas, increasing your blood pressure by reducing potassium levels. That food is liquorice and surprisingly it has killed more than one person. Surprisingly, because you need to eat more than sixty grams of liquorice a day for more than two weeks to experience the bad effects. That's basically an entire bag of Liquorice Allsorts every day for a fortnight. Given my

---

* Eric Wimp endorses this message!

family banned me from ever buying them again (because inexplicably they found them disgusting), I think I am safe from Bertie Bassett's blood pressure boost.

If you are not a fan of bendy yellow fruit, beetroot juice might provide an alternative food-based solution, as I learnt about from a relative desperately consuming it in an attempt to reduce his blood pressure under spousal pressure. As with many of these edible interventions, it sounds a leetle beet tenuous, but in a (albeit relatively small) randomised clinical study, 250 ml of juice a day did indeed reduce blood pressure.[13] The proposed benefit comes from the nitrates in the beetroot, which break down into nitric oxide, the same gas found in a small silver capsule littering a city park near you having released its cargo of laughing gas into some undesirable teenager.

Nitric oxide is a jack-of-all-trades molecule in the human body: immune cells use it to kill bacteria; it signals through the nervous system (hence the laughter) and it prevents platelets from causing clots. It also acts as a potent vasodilator, relaxing muscles by changing their cellular chemistry. Nitric oxide has such importance that the researchers working on it, Furchgott, Ignarro and Murad, won the 1998 Nobel Prize 'for their discoveries concerning nitric oxide as a signalling molecule in the cardiovascular system'. Scientists tried to exploit the beneficial value of nitric oxide with drugs. One drug, called Sildenafil, mimicked some of the nitric oxide signalling pathways in the lab. It failed to alter blood pressure in phase I trials. However, it did produce a surprising side effect: men in the study experienced more erections. This led the tumescent team at Pfizer to repurpose the drug as an erectile dysfunction medicine and market it as Viagra, with peak sales of nearly $2 billion in

2008! I've not asked my relative whether beetroot juice had the same effect.

When beetroots and bananas fail, drugs exist that deliberately target high blood pressure without engorging effects – the antihypertensives. The most common of which are the ACE inhibitors. ACE stands for 'angiotensin-converting enzyme', and you might remember it from the COVID-19 pandemic – SARS-CoV-2 binds to the ACE2 receptor on cells to enter them. When not acting as a convenient entry path for a killer virus, ACE is part of the renin-angiotensin-aldosterone system. Like clotting, this is another complicated biological cascade that appears to have evolved specifically to confuse medical students.* The ACE enzyme activates a protein called angiotensin, triggering processes that increase blood pressure. ACE inhibitors interfere with this pathway; reducing ACE activity lowers blood pressure. Dosage of the drug requires careful calculation, as too much of it leads to dizziness by preventing the blood getting to where needed.

In an excellent example of nominative determinism, John Vane discovered ACE inhibitors; other nominative examples include urology researchers Splatt and Weedon, the astronomer Professor Heavens, Lord Brain who literally wrote the book on neurology and my colleague's husband, A&E physician Dr Will Hurt. ACE inhibitors have a tenuous banana link. The research that discovered them originates in banana plantations, specifically in the pickers who collapsed due to rapidly dropping blood pressure after being bitten by Brazilian pit vipers. A Brazilian scientist, Maurício Rocha e Silva, isolated a compound from viper

---

* See also complement (immunology in-joke).

venom called Bradykinin. In the 1960s, John Vane demonstrated that bradykinin blocked ACE. Chemists at ER Squibb expanded upon the idea, testing more than two thousand compounds to develop a synthetic inhibitor called Captopril. Thanks to the WHO's international nonproprietary names (INN) programme, all ACE inhibitors use the stem -pril; the book of stems contains 220 pages covering every drug class imaginable.[14] Whilst this sounds extraordinarily dull, it is extremely useful because if you find yourself rushed into hospital doctors can know roughly what types of drugs you are on, even if they don't know the specific brand.

GOOD FAT BAD FAT

Elevated blood pressure is part of the risk equation for stroke and other circulatory diseases. Whilst pushing the blood around at a higher pressure has some inherent risks by itself – such as bursting blood vessels, leading to internal bleeding and clotting – there is a synergistic interplay with the critical co-risk factor, blood cholesterol. Cholesterol belongs to a family of biochemicals called lipids. We mainly find it in the membranes that surround our cells, but it also is the starting material for many hormones including cortisol. As with blood pressure, cholesterol has a Goldilocks range, too little being as bad for all-cause mortality as too much. Lacking cholesterol increases the risk of depression and anxiety, proving cheese is good for the soul. But too much cholesterol causes far more problems. Cholesterol is the chemical basis of atheromas, the fatty plaques that either restrict the size of arteries or float off to cause trouble elsewhere. Cholesterol acts a bit like mustard in a good

French dressing,* emulsifying the water and the fat. This stickiness means other things can adhere to it: high cholesterol forms the fat in the fatberg analogy (to the platelets' baby wipes).

Simplistically, cholesterol can be divided into two types – good and bad. Unsurprisingly, you need more good and less bad; something appreciated by the manufacturers of butter alternatives everywhere. What distinguishes the two types? The good type, high-density lipoprotein (HDL), is a mixture of protein and lipids that shuttles around the body, stopping plaques from forming. All other types are bad news; every unit (mmol/l since you asked) increase in non-HDL cholesterol leads to a five percent increase in cardiac disease incidence over twenty years.[15] Cigarettes exacerbate cholesterol damage, part of the cocktail of crap that smokers inhale increases bad cholesterol. Ditto drinking, though this is probably because it kills your liver rather than direct action on the cholesterol.

One way to increase the benefits of HDL (the good cholesterol) is more exercise; and whilst increased exercise is *always* a good thing, the evidence for its exact impact on HDL appears to be quite mixed. Some studies have shown it increases the amount, others that it changes how it works in the body. As with all these things it is complex. The other lifestyle intervention is to change diet. For a long time, eggs were regarded as problematic due to their high cholesterol levels. However, it is not just about reducing cholesterol

---

* Perfect vinaigrette recipe: one part wine vinegar, one part olive oil, quarter part mustard; make it in an old mustard jar. Screw lid tight before shaking. Alternatively, Sam (the book's editor) suggests 3:1 oil to vinegar, with less mustard but add garlic and pepper; however, I would take his dressing with a pinch of salt.

intake, it is about reducing it in combination with other saturated fats. This means that an egg alone is fine, but an egg fried in lard, served on buttered toast with a side of bacon and a deep-fried hash brown is bad. Saturated fats tend to come from animal products, but coconut oil contains sky-high levels, about a third more than butter.

As with blood pressure, when diet fails, we turn to drugs. The most common of which are the statins. Indeed, statins are one of the most commonly prescribed drugs of all in the UK, with nearly ten percent of the whole country taking them. Statins don't affect dietary cholesterol; they affect the amount our bodies make in the first place. Because we need cholesterol to live, we make our own, approximately one gram a day in an average-sized adult. Akira Endo discovered statins whilst looking for antibacterial compounds and, like the microbiologists Selman Waksman, Albert Schatz and Alexander Fleming, had turned to microorganisms as a source of novel drugs. He identified a compound called mevastatin from *Penicillium citrinum*, a fungus related to the yeast that makes penicillin (*P. chrysogenum*) and the one that turns Roquefort cheese blue (*P. roqueforti*). Statins prevent disease incredibly effectively, cutting the risk of heart attacks and strokes by twenty-five percent. As with all drugs, statins can have some side effects, and curiously grapefruit juice can make these worse – it contains a compound called furanocoumarin which blocks the breakdown of statins in the blood. Which is as good a reason as any not to eat grapefruit.

SO, SO COLD

A number of factors left me looking for a quick hypertension win: the beastliness of beetroot juice, the joylessness of unsalted food and my pridefulness preventing me from taking age-appropriate medications. Having so convincingly failed to increase my amount of exercise when writing about the heart, I looked for an alternative to reduce my blood pressure. And what could be more 2020s than immersion in cold water, which proponents suggest has beneficial effects against cardiovascular, obesity and metabolic diseases? Admittedly, these benefits are a bit sketchy.[16] As I found myself in Cornwall at the time of writing, I decided to take the plunge in the sea every day for a week[*] and record the outcomes in an entirely subjective way.

Day 1. Weather rainy and windy; sea temperature ten degrees Celsius. Alone (my daughter refused to join me because of the rain). Length of time in sea – five minutes. It was only this long because I shared the beach with someone else who had clearly just been swimming and I felt ashamed by my weakness.

Day 2. Weather rainier; sea temperature still cold. With daughter for five minutes, but she got out quickly because she only had one sea shoe – the matching one having been lost six months previously. Length of time in sea – ten minutes. I lasted longer because I remembered my wetsuit gloves this time.

Day 3. Weather windier; sea temperature unknown. With family. Length of time in sea – zero minutes. We had

---

[*] In a wetsuit, with boots and gloves. It was the UK, in March. And make sure there is lifeguard supervision – open water swimming is dangerous, kids!

changed beaches to the rougher north Cornish coast to discover waves crashing on the sea wall with enough force to break bones. Whilst internal bleeding probably would lower blood pressure, we decided that discretion was the better part of valour.

Days 4–6. Some sun, not warm. With extended family. Length of time in sea – thirty minutes.

Because I am a proper scientist, I repeated the whole experience a year later – no reproducibility crisis in my research! I confirmed the original findings, the English Channel in late March is bitterly cold, but spending time in it with my family is fun. Sea-swimming did seem to resolve whatever calf injury I had given myself, so score one to the curative properties of cold salty water.

From my studies, it looks like I made some amazing breakthrough in resilience, but being honest, most of the increase in swim length came about because the weather improved and I remembered all the right bits of kit. Having (finally) warmed up, I looked into the delicate balance between the pros and cons of cold water swimming. Cold water swimming is not without risk. Most advice emphasises that cold water can lead to shock, arrythmia and hypothermia; none of which sounds that healthy. One scientific paper stated that whilst cold water has some health benefits for the already healthy, 'there is a risk of death . . . either due to the initial cold shock or progressive decrease in swimming efficiency'.[17] Cold shock is where the body responds to the cold by desperately diverting blood away from the extremities in an attempt to preserve core temperature; not dissimilar to the shock of blood loss. This increases both blood pressure and heart rate and if you are that way inclined can trigger a heart attack. Any water below fifteen degrees

Celsius can cause cold water shock, which basically means most of the water in the UK can kill you. If it doesn't stop your heart, the cold can take your breath away, causing you to gulp in water. In addition, there is the problem of sewage. Right now sewage alerts are being splashed along the south coast of Britain. The beach at which we swam isn't monitored at this time of year, so let's pretend it was OK as I don't want to add gastroenteritis to my list of things to worry about. All told, the levels of sewage in UK waters have reached crisis point (though admittedly they are much cleaner than the nadir days of the seventies and eighties when I was a child). As Chris Whitty, the chief medical officer, said: 'Nobody wants a child to ingest human faeces.'[18] The risks of death or disease by icy water are low, but not zero. The Outdoor Swimming Society (which presumably has some vested interest in people swimming outdoors), estimates that there is less than 1 death per 50,000 swims in England and 1 incidence of sickness per 9,094 swims;[19] but that's still sixty people dying annually – about the same as the number of people who die of HIV in the UK.

What about the benefits? Swimming clearly can give a psychological boost. Without doubt I felt better for doing it, but it is very hard to subtract my being on holiday from any cold-water specific effects. And I think this is the challenge with viewing cold water swimming as a health approach. Any benefit is likely to be marginal and probably combined with other effects. Most outdoor swimming is done in beautiful places, not industrial runoff; being in beautiful places makes you feel better. The beach I swam at is sufficiently beautiful that it recently achieved that irritating status of being written up as 'a destination'. Being there, even without freezing your cojones off, nourishes the soul, which brings the blood pressure down more than anything. There is some anecdotal evidence that cold water swimming reduces the number of respiratory infections but picking apart the detail of how this might work is a mess. A well-written review (a scientific article summarising and evaluating other people's research) reports no measurable changes in immune function – or, to put it another way, the types of small changes being reported in the original studies wouldn't find me popping champagne corks.[20] Cold exposure alone may not be enough; in one study, the addition of breathing exercises had more effect, though again the data is pretty noisy.[21] If you know what you are doing, and you like it, cold-water swimming has benefits. Teasing apart the exact contribution of the cold water from other things such as location, demographics of outdoor swimmers and spending time with others is very tricky.

I will leave the last word about swimming to the Royal National Lifeboat Institution (which knows a thing or two about the dangers of water): 'treat water with respect'.

I have now reached the far shores of stroke and diseases of blood circulation, but not unearthed any unexpected life-preservers. Eating less salt and maybe more bananas can provide some benefits, but the main advice, as with the heart, is to do more exercise. This can come through open water swimming, but plenty of alternatives are available. One thing is for certain: that first plunge into the icy depths will take your breath away. Unfortunately, many other things may do the same, contributing to the next cause of death – lung failure.

CHAPTER 6

A Fate Worse than Breath: Killer Number 3 – Lung Disease

'What oxygen is to the lungs; such is hope to the meaning of life.'

Emil Brunner

Oxygen makes up one-fifth of the air around us and yet heart attacks and strokes kill us by depriving the brain of oxygen. Our bodies are simply too big and complex to directly absorb the vital gas we need from the air, which is why we evolved a complex circulatory system to deliver oxygen where needed. But how does the blood acquire it?

Back to the interactive science. Take in as deep a breath as possible and then time how long you can hold it for. The average is somewhere between thirty and ninety seconds; the record is a remarkable twenty-four minutes and thirty-seven seconds held by Budimir Šobat, a fifty-six-year-old Croatian free-diver. The actor Kate Winslet holds the slightly more niche record for length of breath held on a movie set at seven minutes fifteen seconds,* beating Tom Cruise's earlier record. These records might prove challenging for you to beat, firstly because you are not a movie star,

---

* Proving that not only her heart but also her lungs can go on.

but more importantly because both actors (and Budimir) inhaled pure oxygen before attempting them. The record with inhaled air alone is still pretty staggering at eleven minutes thirty-five (or fifty-four) seconds (depending on who you believe). Sadly, I am unlikely to beat either record; as with cardiac output, our ability to breathe declines with age, from a peak volume of about six litres in the mid-twenties. As a 'fun' challenge see how much of this chapter you can read in one breath; the older of you out there may struggle to get off this page, or even this sentence if you read slowly.

Breathing is a curious act: most of the time we breathe without thinking about it at all, until we do think about it and then it can be quite hard to return to subconscious control. On average, we breathe about twelve times a minute, but this varies according to need. You will no doubt find that while reading this section you become much more conscious of your breath. Our ability to breathe without thinking comes back to the medulla, home of involuntary behaviour. The heart's system of speeding up and slowing down also applies to the lungs. This makes sense, as you don't want your blood being pushed around the body faster if there isn't any more oxygen for it to collect when it passes Go. Whilst it can be temporarily overridden, the autonomous nervous system does eventually reassert itself in the control of breathing – that's why you cannot suffocate yourself just by holding your breath, however much you want a squirrel in a magic chocolate factory.

The composition of gases in the blood determines our breathing rate. Just as a quick reminder to make sure everyone is on the same page: we need oxygen (O_2) to live because it helps us burn sugar to produce energy, creating waste in the form of water and carbon dioxide (CO_2); the process is called respiration. Your body can measure the levels of both vital gases involved in respiration, oxygen and carbon dioxide. CO_2 levels play a key role, determined through changes in the blood pH. CO_2 is conveniently a bit acidic (think fizzy drinks); if its level rises, the blood acidifies. The inference of CO_2 rising is that O_2 has dropped, and the medulla counters the perceived lack of O_2 by increasing the breathing rate. You can test this out: put your head under the duvet and see how it affects your breathing rate (it should increase). The level of CO_2 in exhaled breath may alter the orientation in which couples sleep together in bed. Most people do not sleep facing each other: according to a survey, forty-two percent sleep back to back, thirty-one percent sleep facing the same direction and only four percent face each other.[1*] Inhaling someone else's exhalation speeds up your breathing because exhaled air contains one hundred times more CO_2 than atmospheric air; hence the facing away.

The mechanics remain the same regardless of whether we breathe consciously or unconsciously. Two sets of muscles control inspiration: the diaphragm (a long, thin strip running under the bottom of the ribcage and connected to the breastbone) and the intercostal muscles (the ones between the ribs). The muscle sets work together. The diaphragm pulls downwards, the intercostals outwards, increasing the total volume of the lungs and causing air to rush in through the nose and mouth. Exhalation runs the

---

* The press release didn't say what orientation the remaining twenty-three percent of people slept in, but if I had to guess it would be in separate bedrooms.

same process in reverse; the lungs are squeezed, and the air forced out. Bellows, pipe organs and accordions all work in the same way: increasing volume brings air in, reducing it forces it out. An average adult breathes in half a litre of air each breath; but the lungs don't completely empty and refill each time, and most breaths only replace about ten percent of the air in the lungs.

Air should enter the body through only two routes – the nose and the mouth. It can enter other ways, none of which are good! Furthermore, gaseous air in our lungs is a good thing; air in the bloodstream is a bad thing. Called a gas embolism, the air blocks the passage of blood through the vessel, acting like one of the rogue clots that we encountered in the last chapter. Gas embolisms, known as 'the bends', can occur after deep water scuba diving. They can also come from injecting air directly into the blood vessels, a time-honoured murder method from whodunnits.

The preferred inhalation route is the nose, because it evolved specifically for this purpose; nasal breathing improves the quality of the inhaled air by trapping debris in the nasal hairs (though why nasal hair length increases as we age remains to be determined*) and humidifying it before it gets to the lungs. One curious feature of nose breathing is nasal cycling. This is where one side of the nose becomes congested, whilst the other side clears and vice versa. It is controlled by the same type of erectile tissue found in the clitoris or the penis; which may explain one of the side effects of Viagra, blocked-up noses – a sudden rush of blood

---

* Don't get me started on eyebrows, mine apparently have been possessed by the ghost of Denis Healey.

to the head!* You may have experienced nasal cycling during a cold, when miraculously one nostril becomes clear, but then infuriatingly blocks up again.

Having entered the head, the air then funnels into the trachea (windpipe). At this juncture we encounter a fairly major design flaw – the proximity between the breathing and eating holes. The tube that carries food to the stomach, the oesophagus (sometimes called the gullet) sits right next to the trachea, with hilarious consequences (not really hilarious). We have evolved some protection to avoid inhaling peas: the action of swallowing shuts a trapdoor-like flap of tissue called the epiglottis over the opening to the trachea, keeping the route to the oesophagus clear. Different animals use different systems: the horse epiglottis prevents anything getting from the mouth into the airway, even air, forcing them to be nasal breathers only. The same is true for rabbits, cats and rodents, all obligate nasal breathers; ditto human babies, reducing the risk of inhaling suckled milk.

The colocation of the food pipe and air pipe means that occasionally things go down the wrong one. Air in the food pipe can be embarrassing (or epic depending on your age/mental age), because it causes belching. Food in the air pipe is more serious. Mostly this can be dislodged by coughing or the tried and tested 'whacking someone on the back way harder than required' method. But in some cases, trapped food needs dislodging by abdominal thrusts, better known as the Heimlich manoeuvre. This involves wrapping your arms around a person from behind, jamming your fists in between their ribcage and belly button and then forcefully

---

* It also makes you wonder what Pinocchio was lying about!

tugging up into their diaphragm several times until the trapped food is expelled through the mouth. It's named after the American thoracic surgeon Henry Heimlich who developed the manoeuvre in the 1970s. Having tested his ideas on beagles to whom he had force fed large chunks of beef, he published his landmark work 'Pop Goes the Café Coronary'. Current guidance says do the whacking first, before trying the Heimlich. However, I can attest to the Heimlich manoeuvre being effective, having witnessed it first-hand. At lunch, a school friend inexplicably decided to replicate Henry's doggy dabblings by inhaling an enormous slice of roast beef. He began to turn a slightly odd colour. This led to a debate between two other friends, one medically qualified, one in medical training, as to who would do the deed. The trainee won/lost the argument, the piece of cow reappeared, and my friend returned to normal colour. Whilst saving his life, it did, however, put the rest of us off lunch, so swings and roundabouts.

The trachea is a reinforced tube, with rings of cartilage that prevent it from collapsing under the pressure of inhaled breath. The tube then splits into two, called bronchi, heading for your lungs, followed by further branching and narrowing into bronchioles and finally alveoli, where oxygen molecules can find their way into the bloodstream. Unlike a bellows, the lungs aren't just a massive undifferentiated sack, they have a highly organised structure. If you picture your lungs in your mind's eye, you probably see two symmetrical halves, but the lung is actually made of five lobes, three on the right and two on the left. The heart taking up the space on the left-hand side that would otherwise be lung (Figure 10). In about one in ten thousand people, the organs are reversed, with the heart located on the right-hand side.

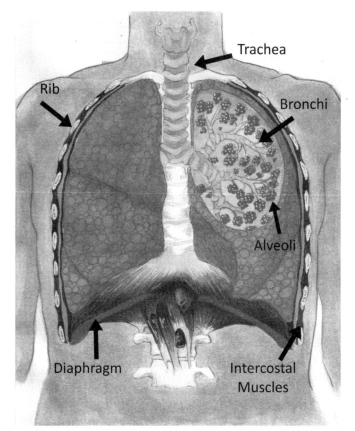

Figure 10. Every breath we take. The lungs are where we exchange good oxygen for the waste product carbon dioxide. They have a very high surface area to enable this, read on to find the oft-quoted sports analogy.

COUGHS AND SNEEZES SPREAD DISEASES

The lungs play a central role in a healthy and happy life. As with the heart, they are quite busy. Breathing about twenty thousand times a day exposes them to a range of noxious substances, and defences have evolved to stop them getting damaged. As well as the larger hairs found in the nose, smaller hair-like projections, called cilia, line the airways. The cilia beat ten times a second in a co-ordinated pulse from the bottom of the lungs to the top, moving foreign particles out of the lungs. Cilia are coated with a fluid called mucus; you will most likely know this as snot or phlegm. Mucus traps objects, moving them to the top of the windpipe to be swallowed. The most common particulate trapped in snot is dust; we inhale fifty billion dust particles an hour, a cocktail of dirt, bacteria and human skin. A study of New York Metro dust identified that ten percent of the inhaled dead skin comes from people's feet – five percent the buttocks.[2] On average you swallow a litre of mucus a day; having a cold can double the volume. As anyone with even the slightest degree of curiosity can attest, there is a rainbow spectrum of snot. Mucus should be colourless but can go from yellow/green (infected), red (fresh blood), brown (old blood) to black (you live in London). We tend to blame bogeys for blocking our noses, but it isn't the only culprit: the influx of immune cells fighting the infection clogs up the airways and swollen nasal erectile tissue also plays a part.

In addition to sticky mucus and cilia beats, our bodies evolved more explosive mechanisms to remove foreign objects from the airways – coughs and sneezes. Both are involuntary actions; sneezes clear the nose, coughs the lower airways. Coughing and sneezes do in fact spread diseases. One lasting impact of the COVID-19 pandemic is that more people now use the vampire method, coughing into the crook of their elbow rather than their hand. Though, as I have discovered, this comes with the unpleasant downside of occasionally coughing up gloop straight onto your shirt.

Sneezing is initiated by particles landing on the surface of the nose, triggering the release of histamine. Histamine release can also follow irritation or damage, which explains why pulling out lengthy nose hairs makes you sneeze. Histamine activates nerves in the nose that run up to the brain and back to the muscles lining the airway and chest, resulting in an explosive expulsion of air, up to one hundred miles per hour. Immune cells in the nose also produce histamine in response to allergens such as pollen, explaining why antihistamines (drugs that block their action) can treat hay fever. There are three things worth exploring about sneezes: one truth, one part-truth and one lie. The truth is the connection between sneezing and staring at the sun; this has rather tenuously been called the autosomal dominant compulsive helio-ophthalmic outbursts of sneezing (ACHOO).[3*] Sun-sneezes are probably triggered by the proximity of the optical (seeing) nerves and the trigeminal (sneezing) nerves, so stimulation of one might fire the other, confusing your medulla into sneezing. The part-truth is that a sneeze is a fractional amount of an orgasm. Whilst the nose does contain erectile tissue, there is no connection between the two; but sneezing does release endorphins, meaning there is some pleasure associated with them. The lie is that your eyeballs pop out if you sneeze with your eyes open. I can find no explanation linking dark chocolate consumption with sneezing, though it is an irritating tell when I am trying to snaffle my wife's secret stash.

---

* Forced, tenuous acronyms are a job hazard of being a scientist, to such an extent that the *BMJ* did a 'humorous' article on it called 'SearCh for humourIstic and Extravagant acroNyms and Thoroughly Inappropriate names For Important Clinical trials (SCIENTIFIC): qualitative and quantitative systematic study'. The authors noted an increase in number and a decrease in quality of acronyms over time. The many examples of bad acronyms are too painful to include here.

The other way to expel things from your airways is coughing. To list the causes of an acute cough by increasing severity, one goes from the inhalation of irritants and upper respiratory infection before accelerating into pulmonary embolism, collapsed lung and heart failure. Don't let this list alarm you too much; colds/irritants cause the vast majority of coughs, not imminent heart failure. Like sneezing, coughing has sophisticated nervous system control, triggered by receptors in the airway. These receptors recognise both mechanical (things stuck) and chemical (irritant compounds) stimuli. One example is the capsaicin receptor, which registers heat, both physical and jalapeño-induced. The mechanical sensors for coughing not only reside in the airways, they are found also in the ears – which explains why shoving a cotton bud in your ear can make you cough. Hopefully by now you can guess which part of the brain the receptors trigger?* Though in this case there is some ability to override the autonomic system and suppress the cough, which is handy when at the world snooker finals at the Crucible in Sheffield.

On stimulation, the medulla fires back to the muscles in the chest. Coughing has three phases: inspiratory – a big intake of breath; compression – pressurising the lungs in the air by closing the epiglottis (the flap that stops you accidentally inhaling peas) and contracting the intercostal muscles; and finally, expiratory – when the epiglottis opens, expelling the air at force. Acute coughs can be irritating but normally pass, with or without medicine. A meta-analysis of thirty studies revealed that treating coughs with over-the-counter cold remedies may provide psychological relief but

* The medulla of course. Neuroanatomy 101: if in doubt it's the medulla.

probably have little net benefit.[4] A chronic cough, defined as one that lasts eight weeks or longer, is a bigger problem. It affects about ten percent of the general population, with a higher frequency in women and people who take ACE inhibitors. Chronic coughing can be problematic for both sociological reasons (lack of sleep, isolation, incontinence) and physical reasons including cough syncope, which is where you feel faint or pass out after coughing.

Whilst coughs and sneezes clearly serve a purpose, the other involuntary explosive airway event – hiccups – is a bit of an outlier. In case you don't know what hiccups are, Daniel Howes, professor of emergency medicine, Queen's University Ontario in Canada, described hiccups as 'an onomatopoeic name that comes from the sound made by the abrupt closure of the vocal cords approximately thirty-five milliseconds after the forceful contraction of the respiratory muscles'.[5] But since describing hiccups to someone who has never had them is essentially the same as explaining colour to someone who has never seen it, I am going to settle for saying 'you know it's a hiccup' and assume everyone hiccups. The scientific name is apparently a singultus; I would like to suggest that even fewer people have used this word than have never hiccupped.

No one really knows why we hiccup. Singulti* probably relate to early life development, being much more frequent in newborns. Other suggested reasons for hiccups include: training respiratory muscles in the womb, clearing meconium (inhaled baby faeces) immediately after birth, acting as a remnant of the gills and helping move food trapped in

---

* *Reprehendo sicco meus artes Latinae* (trust me, this is a good joke – but you'll probably need Google Translate; I did).

the windpipe. In his paper, Howes suggests that they, like burps, serve as a way to remove air from the stomach. However, they are considerably less fun, and it is much harder to hiccup the alphabet, so I am not convinced by this argument. As to curing them, frankly we are no wiser about how to stop them than as to why they start. Apparently, researchers have developed a breakthrough device called the FISST (forced inspiratory suction and swallow tool), basically a straw, with a reported ninety-two percent success rate in stopping hiccups.[6] The slight problem is that the study lacked a control group, in other words the hiccups might have just stopped anyway. Dr Rhys Thomas, a consultant neurologist, said, 'I think this is a solution to a problem that nobody has been asking for,'[7] which is exactly the kind of feedback my research proposals get!

MEASURING MY LUNGS

If you strip away the rest of the lung tissue, you end up with a fine filigree of tiny vessels closely intertwined with blood capillaries. As with most of the body, form follows function. The lungs have an extremely high surface area, maximising the amount of oxygen that can pass through them and into the blood. The total area of one lung is about the size of a tennis court;[*] if you put the airways together, they stretch fifteen hundred miles (ten times the length of Wales). The volume of the lungs is more important than the area; their structure is vital to how they maintain life. One way to calculate the volume of the lungs is to fill them up with water, which is clearly terminal and not a sensible way of

---

[*] Of course it is, tennis courts being another important SCI unit.

assessing function. More practically, your lungs' capacity can be estimated through spirometry: from the Latin word *spirare*, 'to breathe' – the same root word as for 'respiration', 'perspiration' (breathing through) and 'inspiration' (breathed upon, by a muse). As part of work health screening, I have records of my lung function dating back to 2007. I thought I ought to get it reassessed.

To do this I asked for help from Dr Eva Fiorenzo, clinical research fellow at St Mary's hospital, London. I had been working with Eva on the spuriously named MAGIC (Metformin and Airway Glucose In COPD) study which surprisingly looks at airway glucose in COPD (more of which later). Whilst we had worked together for eighteen months, thanks to COVID and remote working we had never actually met in person until the day of the lung function test, meaning I got to kill two birds with one stone: discuss the project and get my lungs tested. We met at the Imperial College Respiratory Research Unit, housed in the stable block at St Mary's. For clarity, the horses are long gone, and they actually belonged to the Great Western Railway at Paddington station next door, not the hospital. Remarkably the stable housed six hundred horses, necessitating a two-storey stable with a gentle ramp running up to the first floor, and slatted bricks to let the wee run away. As part of the modern NHS estate, very little has changed: slightly fewer horses, slightly more urine. Before her research placement Fiorenzo specialised as an accident and emergency doctor, but admits that the pace of research is somewhat more conducive to normal life than dealing with road traffic accidents at two in the morning. We discussed research's potential magnitude of good: as a hospital doctor you save the people in front of you that day; if research works you can change the outcome for thousands, even millions, of people – which is of course why the government should fund even more of it, especially the vital work in my lab.

Last time I had undertaken spirometry, I blew into a cardboard tube attached to the spirometer, but progress being what it is, the cardboard has been replaced with a single-use plastic tube that goes straight into landfill. I took a deep breath and then blew as hard as possible for as long as possible. This turned out to be not very hard at all! I sounded like a wheezy Thomas the Tank Engine pulling into the sidings for the last time post Beeching. I had a second attempt, and this time I got to watch the screen as an incentive to do better. I didn't. Still, third time's the charm as they say. Or in my case not. I then got my results printed; the spirometer, despite being shiny and new, was attached to a charmingly old-fashioned dot matrix printer.

My experience of spirometry with Fiorenzo was considerably less emotional than the last time I had my lungs investigated in the name of science. In that study, we aimed to measure the chemical contents of the airways.[8] To do this, we needed samples from the upper and lower airways. Getting my upper airways sampled was easy: I put a small piece of fancy blotting paper up my nose. Lower airways less so. To get a sample from my lungs, I had to have bronchoscopy. This involves running a camera on a tube (with a grabber attached) down the trachea into the lungs. Respiratory physicians routinely use this to detect physical abnormalities in the airways. Bronchoscopy belongs to a wider family called endoscopy, which includes gastroscopy (camera inserted via the mouth), cystoscopy (camera inserted via the penis), colposcopy (camera inserted via the vagina) and colonoscopy (camera inserted via the anus). This goes to show that if they can find a hole in your body, at some point a doctor will probably want to stick a camera in it. Sometimes they might even make an extra hole to look. I've been 'lucky' enough to collect three of the endoscopy set.[*] Bronchoscopy was mid-league; I went across to the hospital research unit, had a quick chat with the study lead, Dr Patrick Mallia, who then sprayed some burnt banana-flavoured painkiller on my voice box before administering the main anaesthetic – flunitrazepam. You may have heard of this by its trade name Rohypnol, in the context of being a 'date-rape' drug. It is potent stuff. One minute I was chatting to Patrick, next thing I knew I was also chatting to Patrick . . . the only difference being two hours had passed, and I was in a different room.

Gastroscopy tops my worst endoscopy experience list. I opted out of anaesthesia because I needed to be able to drive and pick up my kids from school afterwards. Big mistake. The camera passes the gag reflex on the way down, and if you don't know where this is, shove your fingers down your throat and see what happens.[†] The placement of the camera meant for the next twenty minutes I was desperately trying not to throw up, but of course couldn't because there was a camera in there. At one point the doctor said to me – try not to gag, it upsets the camera, to which I would have replied, 'You're joking aren't you,' except again I couldn't because I had a camera down my throat. I had learnt my lesson for the colonoscopy and ticked the 'all drugs' box and told the kids to walk home. The only downside to the colonoscopy was that since sweetcorn (and other foods) can

---

[*] And whilst I pitched this book to contain self-experimentation, there is no book deal large enough to have cystoscopy for the sake of a good anecdote.

[†] Actually don't, unless you want to vomit all over yourself and my book.

look a bit like cancerous polyps you need to both starve yourself and clear the system out with some extraordinarily potent laxatives. Leading to twelve hours of toilet-bound misery with nothing but thin broth and Lucozade to drink.

DOCTOR, GIVE ME THE NEWS

Armed with my new measurements, I met up with Dr Hugo Farne, a respiratory consultant specialising in asthma to see what they meant and how he uses these numbers to diagnose and manage respiratory conditions. We had first met at another friend's wedding – turns out weddings are an important source of science expertise. Hugo is tall and was casually dressed in jeans and a T-shirt, suits apparently for the previous generation of consultants (admittedly he had no patient-facing work the week we caught up). Also, for some reason he had a pair of wire cutters with him, I thought it best not to ask why. We started off by discussing the various types of lung measurement. The simplest is the peak flow meter, which uses a little arrow to point how hard you can blow out. This is often used at home by people with asthma to identify changes in their lungs.

The spirometer is more sophisticated because it can measure both the volume of air exhaled and the speed at which it leaves the lungs. Spirometry returns two important types of data. The first is the Forced Expiratory Volume (FEV1), the amount of air exhaled in one second. The second is the Forced Vital Capacity (FVC), the amount of air blown out after a full inspiration. I combined my new data with that from previous measurements and, because I am a scientist and love a good graph, plotted the data over time (figure 11). Looking at these plots, I couldn't escape the conclusion that it is all headed downwards.

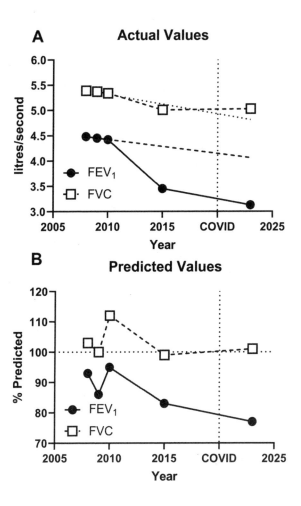

Figure 11. John's worsening lung function. Airway data recorded by spirometry. A. Actual data recorded. B. Predicted values based on my age, sex, height and weight.

To get a more detailed evaluation, I showed the data to Farne. His first comment was 'Oh' . . . and not a reassuring 'Oh' either. He went through the data to me. 'Spirometry and in particular FEV_1 is helpful because it is easy to do and it has been in use for a long time,' explains Farne. Lung function measurements date back to the late seventeenth century, and the earliest approaches involved blowing into a tube full of water to see how much could be moved. John Hutchison refined this approach in the 1830s on behalf of the Britannia Life Assurance Company, replacing the tube with a bucket.[9] He mainly measured whether or not people had tuberculosis (the predominant lung disease of the time) and collected data on over four thousand subjects including 'sailors, fire-fighters, policemen, paupers, artisans, soldiers, printers, draymen, pugilists, and wrestlers, giants and dwarfs, gentlemen, girls, and diseased cases'. Whilst Farne sees fewer wrestlers or pugilists than his predecessor, he does use one of the measures that Hutchison coined – vital capacity (FVC), the total amount of air breathed out. Vital capacity is not quite the same as the total volume of the lungs as they retain some residual volume (since the lungs are not completely squeezed empty each breath).

We returned to the most important subject – what's going on with my lungs. In the graphs, you can see the actual values in the left-hand panels and the predicted ones on the right. As with all of our bodily systems, lung function undergoes an inevitable decline with age.[10] The good news was that my FVC is pretty much as predicted. However, and what prompted the 'Oh', my FEV_1 was much lower than predicted. Initially I thought this a pandemic after-effect. Even though I didn't become very sick with COVID, it came with a long tail of recovery when I felt quite puffed walking and disappointingly failed to sing at full gusto at a friend's wedding. Fortunately, my singing volume has recovered, though I suspect my opportunities to headline Glyndebourne might lie well behind me.* My lungs' biggest plunge downwards occurred between 2010 and 2015, before SARS-CoV-2 had even infected its first bat/pangolin let alone made the jump into people. Given the drop in my FEV_1 happened more than a decade ago and I can barely remember writing the last paragraph ten minutes ago, I have no clue about when or why this decline happened. Especially as in the time window between 2010 and 2015 I moved out of London and into the suburbs, which should improve airway function by reducing pollution.

My own wheezy lungs aside, how do respiratory physicians like Farne use airway measurements, and what diseases are they trying to diagnose?

BAD BREATH

The predominant lung disease in ageing is called chronic obstructive pulmonary disease (COPD). Over three million people in the UK live with COPD. And rates of COPD are increasing, making it the third-leading cause of death worldwide in 2019, causing nearly three million deaths. Lungs with COPD progressively deteriorate, impacting day-to-day activities. The two most common presentations are chronic bronchitis and emphysema. Bronchitis is driven by inflammation of the lower airways (the penultimate

---

* Not that singing at Glyndebourne was ever really on the cards – reminding me of the famous joke about being able to play a piano after an operation.

branching of the tubes); emphysema is characterised by destruction of the alveoli (the little sacks at the end of the tubes). COPD used to be split into the alliterative categories of blue bloaters and pink puffers. People with chronic bronchitis were named blue bloaters, because breathing difficulties decreased oxygen in the body, leading to a bluish tint on the skin, and they were also more likely to be obese. Those with emphysema were pink puffers because they had to take short, fast breaths, making them pink in the face.

COPD is a disease of the immune system. As a major interface between the body and the outside world, the lungs are lined with a filigree of immune cells protecting them from infection and inhaled material. Unwanted immune responses can be dangerous, and a large number of checks and balances exist to stop them. Even at the peak of the response to an infection, the immune system triggers pathways to shut itself down in what is called the resolution phase. An important aspect of resolution during infection is that the immune system clears away the infectious agent, removing the fuel from the fire. In COPD, and other inflammatory lung diseases, the cells of the immune system remain in a non-resolving state of activation, damaging the delicate latticework of the lungs. This is because the thing triggering the immune response (be it cigarette smoke, industrial pollutant, black mould or, as we shall see, pigeon excrement) is inhaled again and again, keeping the cells in a constant state of emergency. In the absence of pathogens to attack, the immune system turns on the body.

Once immune-mediated damage gets going in COPD, it becomes almost impossible to stop. The decline of the lungs accelerates. Because of its complexity and heterogeneity, various systems have been proposed to categorise the severity of COPD, none of which is perfect. They mostly use the FEV_1, which can get very low indeed. Under the GOLD (Global initiative for Obstructive Lung Disease) scoring system, the FEV_1 of people with mild COPD is eighty percent of what it should be, declining to less than thirty percent capacity in very severe COPD. Losing seventy percent lung capacity is grim; each breath providing only two-thirds of the oxygen required. And lung function doesn't just decline gently in COPD, it jags down in acute bursts called exacerbations. These are predominantly caused by viral infections, which further stoke the immune system to damage the lungs.

What sets the lungs on this self-destructive path? The major trigger for COPD is smoking, accounting for seventy percent of cases in high-income countries and thirty to forty percent in low-income countries, where household air pollution (from cooking fires) also makes a significant contribution. Smoking is bad news. It contributes to heart disease and stroke (through blocking the arteries), asthma, pregnancy failures, low birth weight babies, diabetes and blindness. It probably kills more people through COPD than cancer. Worldwide in 2017, 3.3 million people died of tobacco-related deaths with 1.5 million dying of chronic respiratory conditions, 1.2 million from cancer and 600,000 from increased susceptibility to infections.[11] Every year smoking kills as many people as the entire population of Wales; that's a lot of close harmony choirs and rugby teams. We will return to the terrors of tobacco, but for now let's look at the other major cause of respiratory problems – air pollution.

DIRTY BIRDS

We inhale air pollution outdoors from car exhaust, industrial waste or forest fires; indoors it can come from stoves and fireplaces. Air pollution causes seven million premature deaths a year. The WHO estimates that ninety-nine percent of the global population, so basically everyone living, breathes in air that falls short of their good quality guidelines. The air we breathe contains a heady cocktail of pollutants that damage our lungs; some are gases (NO_2, O_3, CO, SO_2), some are toxic chemicals like lead and some fall into the broader category of PM2.5, a mixture of minuscule but 'orrible, lung-damaging stuff.

At the time of writing (5 June 2024), Delhi held the undesirable 'most polluted city' crown; having knocked Lahore off its top spot in 2023 – though this is probably one India-Pakistan competition neither side wants to win. The chart is in constant flux, but Kampala, Kinshasa, Sao Paolo, Lima and Wuhan all vie for that coveted 'top' spot. The cleanest city crown belonged to Salt Lake City and the cleanest country, Guam. London found itself mid-table, doing some gentle damage but not entirely wrecking my lungs. If we cast the net of history back far enough, London was once one of the great air pollution hotspots. To understand more about air pollution in the Big Smoke, I spoke to Dr Gary Fuller, senior lecturer in air quality measurement at Imperial and author of *The Invisible Killer: The Rising Global Threat of Air Pollution – and How We Can Fight Back*. I caught up with him over the internet, but in the background could see the West London skyline and importantly the air monitoring devices he uses to sample London's skies. Fuller has long championed improving the air we breathe, which you think should be a no-brainer, but as I was to find out this has become increasingly politicised.

We started by talking about the history of air pollution. Attempts to clean up London have been made since 1306 when King Edward I prohibited artificers from using sea coal. One of the first people to write about the negative effects of air pollution was John Evelyn in the excellently named *Fumifugium*. Coal smoke polluted London's skies for much of its history, worsening through the Victorian period, perfect for Moriarty to hide in (crime significantly increased during smogs). The pollution became so bad that trains couldn't run and factories stopped. The government of the time took the first step in the right direction with the Public Health Act for London, 1891. But the big turning point according to Fuller was the 1952 Great Smog, which killed at least four thousand, but probably closer to twelve thousand, people. This led to the Clean Air Acts of the 1950s and 60s, accelerating the improvement, and by the 2020s London had twenty times less pollution in the air than Delhi. As far as records go, London's levels today have reached an all-time low – though presumably the air of Londinium didn't do much damage to the lungs of the Regnenses or Trinovantes, but their cooking fires probably did.

Whilst London's air has improved, it still isn't good, Fuller explains that there is actually no safe threshold for air pollution – all inhaled particulates cause damage. He was part of the team that measured the rate of change in London's air pollution during Boris Johnson's mayoralty; based on their model, it would have taken 193 years to get below the EU safe level of pollution, compared to 20 years for Paris![12] This led Mayor Johnson to announce the ULEZ (Ultra Low Emission Zone). It has been extraordinarily effective, leading to a forty-four percent reduction in traffic pollution in central London, 'I've been measuring air pollution in London for thirty years now,' Fuller told me, 'and I can't really think of anything that's produced such a dramatic change.' Johnson's successor Sadiq Khan rolled out the ULEZ to inner London (bounded by the North and South Circular) and then extended it to all London boroughs. Sadly, it has found itself caught up in the libertarian anti-evidence movement, generating a lot of resistance (often inexplicably printed on yellow signboards, in a confused mess of anti-ness). Some people argue that the ULEZ could be considered regressive because it targets older, dirtier cars owned by people with lower incomes. But it is more nuanced than that. As the most deprived areas of London (those in the east – downstream of the prevailing winds) experience the worst air pollution, people on lower incomes stand more to gain.[13] The most recent report (2024) demonstrated a dramatic decrease in levels of air pollution, with London cutting road pollution faster than the rest of the UK. Which is great news for the light peppered moth (less so the dark form) and of course the nine million people who live and work in Greater London.

As someone working in vaccines, I had a lot of sympathy for the unexpected wave of nonsense that Fuller now has to deal with – he says about half of his emails just entail people writing, 'F*** YOU!' The vehemence of the response caught him by surprise, as it runs in the face of the extraordinary weight of evidence demonstrating the damage caused by air pollution. Anti-ULEZ shares a lot of parallels with Anti-Vax: in the 'good old days' of chewing, rather than inhaling, pea-soup thick London air and children dying of consumption, pollution and pestilence were visible, making counter-measures more popular. When the threat becomes less visible (but equally real), preventing it becomes harder.

Air pollution is bad because of the sensitivity of your lungs to repetitive inhalation of tiny things. But pollution isn't the only thing people repeatedly inhale that damages them. One of the more unpleasant lung diseases is silicosis, sometimes called pneumonoultramicroscopicsilicovolcano-coniosis, making it the longest word in most dictionaries. Some people argue that the longest word is the amino acid composition of the protein titin (the largest known protein). This is nonsense (to me at least), as no one ever refers to a protein by its amino acid sequence. To be fair the long form of silicosis was made up specifically to be the longest word (yes, I know all words are made up). Another sesquipeda-lian stalwart, antidisestablishmentarianism, has been dropped by many dictionaries because no one ever really uses it, except as an example of a long word. Unsurprisingly, the German language wins the battle of the mega words. However, *rindfleischetikettierungsüberwachung-saufgabenübertragungsgesetz*, meaning the 'law concerning the delegation of duties for the supervision of cattle mark-ing and the labelling of beef' found itself dethroned after the BSE crisis abated. It has been replaced by *kraft-fahrzeughaftpflichtversicherung* (motor vehicle indemnity insurance). Even more unsurprisingly, Jacob Rees-Mogg holds the record for the longest word in *Hansard*, the parlia-mentary record, with 'floccinaucinihilipilification'.

Silicosis is mostly associated with building work. It is horrible. Small particles of silicon overwhelm the ability of the lungs to clean themselves. When material gets past the cilia and mucus, macrophages digest and destroy it. However, silica is indestructible by the body; to prevent further damage, the macrophages activate an alternate approach to contain it. They release a lot of inflammatory signals instructing the surrounding cells to contain the material, a bit like a scar. Scars in the lung reduce the surface area available for the oxygen to pass through. Eventually the capacity of the lungs to do their job completely fails. Silicosis belongs to a larger family of unpleasant ways to screw up your lungs at work, which can be broadly categorised under pneumoconiosis or lung fibrosis. Asbestosis falls in the same category; the damage comes from inhalation of the fibres – but provided asbestos sheets remain intact it is safe. I would, however, strongly recommend against the decision made by the producers of *The Wizard of Oz* who used asbestos fibre to mimic snow, Glinda not being as good as she thought.* The miner's lung variant of pneumoconiosis comes in many grim flavours depending upon what you are digging up and inhaling, for example siderosis (iron), aluminosis (aluminium), black lung (coal) and stannosis (tin). I am from a Cornish mining family and one of the best things my ancestors did was get out of the pits in the 1800s. As much as BBC's *Poldark* glamorises it, working in a Cornish tin mine was grim.

You can also damage your lungs through your hobbies. Bagpiping can kill, and surprisingly not from the piper being beaten to death by their own strangled cat noisemaking 'instrument'. Fungal infections cause piping fatalities. Blowing into a tartan patterned bag not only produces a dirgeful sound, it also deposits bacteria and fungus there

---

* *The Wizard of Oz* was a somewhat cursed production: as well as asbestos snow, the Wicked Witch was set on fire and the Tinman was poisoned by his makeup – pure aluminium dust painted directly onto his skin. Oh My!

within. The bag part of the instrument is called the bag (I thought it might have a more technical name, but nae) and traditionally was made of leather. Leather bags need seasoning, keeping them supple, the treatment also simultaneously killing any fungus they contained. Modern rubber bagpipe bags don't need seasoning, and so are left to fester, slowly damaging the lungs. There is a string of other diseases where a combination of fungus and allergy damage the airways, grouped under the term hypersensitivity pneumonitis, including: pigeon fancier's lung, cheese worker's lung and, least appealingly of all, hot tub lung. Having hired an inflatable hot tub during the COVID lockdown as a way to relieve the endless tedium, I can easily imagine how that happens; after two days it had become six thousand litres of skin cells, bacteria and other guff. Some of the underlying pathology of these conditions reflects silicosis. The hardworking airway macrophages react to the inhaled fungi causing inflammation in the lung tissue. The condition deteriorates over time because the immune cells become trained to recognise the fungus more rapidly and more aggressively. The risk levels are extraordinarily high (if you breed birds), and remarkably the UK alone has sixty thousand pigeon fanciers. Data from the 1970s suggest one in five get the condition;[14] in terms of risk that puts it up there with horse-riding (forty-six percent chance of head injury) and makes it more dangerous than climbing Everest (ten percent fatality), going into space (three percent fatality) and even base jumping (one death every six thousand jumps).*[15] And given the best way to

---

* These are slightly mismatched comparisons, but one in five is crazy high: it is one thousand times higher than deaths from heroin in Scotland.

reduce the disease is to reduce antigen exposure, there is a simple fix: sell the pigeons and take up base jumping.

Within the category of repeat inhalation, we do need to consider vapes, but I am going to park that till next chapter alongside smoking cigarettes.

ASTHMA

There is another chronic airway condition driven by misfiring immunity: asthma. Approximately five percent of the world's population lives with some form of asthma, with greater prevalence in high-income countries; it affects approximately eight million people in the UK. The word 'asthma' comes from the Greek for 'panting'. As with the other airway conditions, asthma isn't just one thing; it is actually so heterogeneous that a diagnosis of 'asthma' becomes basically meaningless. As a progressively deteriorating lung disease, asthma shares a number of similarities to COPD. It is characterised by unresolved inflammation in the lungs; leading to constriction of the small airways or increases in mucus, both of which make the effort of breathing harder. Chronic lung damage leads to the replacement of flexible airway tissue with more rigid scar tissue, permanently reducing the ability to draw breath. And although we may think of asthma as a childhood condition, it persists into adulthood and may set the stage for other diseases such as COPD.

Unlike COPD, where we know the cause in the majority of cases (smoking), it is much less clear what causes asthma. The forensic challenge is in identifying a transient (possibly unnoticed) event in the past that programmed the lungs into an inflamed-asthmatic state – the butterfly wing-flap that eventually led to a hurricane. One such transient event that can lead to asthma is severe infection with a virus called RSV during infancy. Lucy Mosscrop, a PhD student in my group, is investigating ways we can reduce RSV infection alongside Dr Chubicka Thomas. Asthma has been on the rise in the last few decades, and one suggested reason is called the 'hygiene hypothesis'. This suggests that when the environment in which children are raised gets cleaner, the incidence of asthma increases; though the mechanism is, as yet, unclear.

Once established, asthma can be characterised by acute attacks, leaving the person gasping for breath. Asthma attacks kill three people a day in the UK. The extensive list of asthma attack triggers includes thunder-storms, infections, food, alcohol, emotions, pollen, animals and even coitus (so always keep a blue puffer on hand for safe sex). Triggers subtly differ from causes: causes are the underlying reasons why someone has asthma in the first place; triggers are the things that make it harder to breathe.

INHALER

Once airway disease takes hold you can never restore the lung function, but a range of drugs can ameliorate the symptoms. Much of the work testing these treatments was done by Professor Sir Peter Barnes KBE FRS, who by any measure is an extraordinarily eminent scientist. Just to give one metric – the importance of one's scientific output can be measured by the number of times other researchers cite it in their own work, and for reference ninety percent of papers never get cited again! Barnes' papers have been cited cumulatively 259,716 times.* Accordingly, I spoke to him with some trepidation. Fortunately, much of the tension dissipated because his video conferencing software made him sound a bit like Donald Duck. Having turned it off and on again, we could have a more serious conversation, and it proved an absolute goldmine, given he actually wrote the book on lung disease. We talked about COPD, asthma and the two main treatments currently used: bronchodilators and inhaled corticosteroids.

Early attempts to treat asthma targeted the main symptom – bronchoconstriction (tightening of airways). Doctors tried a range of agents to open up airways (called bronchodilators); remarkably one of the treatments was smoking 'asthma cigarettes' that contained leaves from jimsonweed (*Datura stramonium*), a picture of which by Georgia O'Keefe hangs on my study wall.† This plant contains alkaloids that block the firing of nerves that constrict the airways and thus open them back up again. The twentieth century saw the development of more targeted approaches. Sir David Jack FRS, born in Markinch, a small Scottish village, played a key role. Jack developed an extraordinary range of drugs in his career, including the anti-ulcer drug Zantac, at one point the biggest selling drug in the world, making over $1 billion a year. Zantac was a gateway drug for me: I did a work placement at Glaxo in sixth form, quantifying Ranitidine (its formal name) in human faeces – the glamour of research never faded. Jack developed the first bronchodilator, salbutamol sold as Ventolin – the ubiquitous blue puffer.

---

* My own citation count is much more modest at 7,828 (not that anyone's counting: updated to 7,953 in edit; 8306 in proof, every cite counts).
† Sadly, a reproduction; academia doesn't pay that well!

Prior to Jack's work, derivatives of adrenaline, the 'fight or flight' hormone, had been used to control asthma; these had been somewhat effective because, amongst its other functions, adrenaline opens airways, allowing the inhalation of more oxygen. However, as seen in Chapter 4, adrenaline also accelerates your heartbeat. Early bronchodilators stimulated the whole adrenal response in an unconstrained way and had to be withdrawn because they caused a wave of deaths. Jack managed to isolate the airway-opening part of the drug without the heart acceleration, because different organs use different adrenaline receptors. Airways use the β2-adrenergic receptor, the heart uses the β1 variant. The drug Jack developed is an adrenaline derivative, but only acts on the β2 airway receptor.

The other drugs used are inhaled corticosteroids (ICS). These synthetically mimic cortisol, the other 'fight or flight' signal we encountered earlier in the context of the heart and stress. This may seem like a strange choice for the reduction of airway disease, but as well as firing up the body, cortisol is anti-inflammatory. Corticosteroids shut down the inflammatory cascade by binding to a protein called the glucocorticoid receptor. But corticosteroids cause a wide range of effects, which limited their utility when taken as pills – they often did more harm than good. David Jack pioneered the shift from oral to inhaled corticosteroids, thereby concentrating the effects of cortisol in the place needed, reducing-off-target side effects.

The two-drug combination (Ventolin plus ICS) is highly effective in asthma; when used together they can reduce severe asthma attacks by ninety percent. They used to be administered in two separate inhalers, brown and blue. The brown ICS inhaler had to be taken daily regardless of symptoms, the blue one during attacks, and people forgot to take their ICS, losing the bonus effects of the brown. A simple behavioural nudge offered the solution. Combine the two drugs in one puffer. People with asthma are more likely to use the blue puffer (the bronchodilator) because it gives immediate relief from symptoms, and sneaking some ICS into it means they get both hits at once; like a snooker break – adding a tricky brown to an easy blue.

Whilst there are a number of old (and new) treatments for asthma, the outlook for COPD is less rosy. COPD is characterised by lungs embarking on a pathway to destruction – the drugs merely reduce symptoms, and no drug to date can reverse lung deterioration. As Barnes put it: 'Asthma and COPD are completely different; asthma is largely steroid-susceptible, where COPD is predominantly steroid-resistant.' The only drugs that work are the long-acting bronchodilators. However, 'the current generation bronchodilators are basically as good as they are going to get', said Barnes. It should be a priority area to research new COPD drugs to act on cause not just symptoms. The investment in COPD does not mirror its deadly impact. COPD is the third-most common cause of death in the UK but by far the poor cousin of the chronic diseases. The charity Asthma (and Lung) UK received £15 million in income in 2022 and spent £3 million on research; by comparison Cancer Research UK received £640 million and spent £443 million. This lack of charity support is matched by a similar lack of funding from governments or the pharmaceutical industry. Sadly, if you get COPD, your options are limited. Barnes puts it simply: 'It's a shocking state of affairs.'

BREATHE EASY

In the absence of wonder drugs, what can be done to reduce damage to our lungs? Not breathing in noxious agents in the first place is a good start. And if you are currently inhaling bad things, stop! Easy to say, but I cannot stress how far giving up smoking could extend your life.

Likewise, if you find yourself in a position where you inhale the same rubbish repeatedly, find ways to mitigate the effects. At work, health and safety laws should protect you: areas should be ventilated, and you should get protective equipment to stop lasting damage. If it is a hobby (like pigeon fancying), read up about how to protect yourself – remember the odds are really not in your favour. If you are one of the sixty thousand people in the UK who insist on keeping pigeons, get a proper ventilator to clean out your pigeon loft. More generally, increase household ventilation, get out into the countryside, breathe fresh air. Again, easy to write, not always easy to do. Returning to WHO figures, ninety-nine percent of us inhale air below the acceptable pollution level. Change can happen. Support politicians who are reducing air pollution. Read about the positive impacts of low/no drive regions before getting cross about them. Replace your car with something less polluting, drive less, walk more. Plant trees (whilst the London Plane tree is extolled for its air cleansing virtue, the silver birch, yew and elder all score higher for PM removal[16]).

FORGET ABOUT BIKES

But let's return to the important subject, me and my slightly duff lungs. Could I use any 'hacks' to improve my lung function? As with heart health, exercise offers the simplest path. Improving lung function through exercise is slightly cheating, because some of the benefit felt comes because the rest of the body gets more efficient at using oxygen, therefore reducing the load on the lungs. So, yes, of course, exercise is beneficial, but it isn't solely improving the lungs.

It's also important not to overdo it. Elite athletes can damage their lungs through repetitive inhalation; exercise-induced bronchoconstriction occurs in fifty-five percent of elite athletes.[17] Their repeat inhalation of cold, dry air is problematic – the speed at which air needs to be inhaled in athletes means it fails to warm up properly. Specific sports add a chemical note to this damage making it worse; for instance, swimmers inhale chlorine* and ice skaters inhale nitric oxide and tiny bits of ice. This briefly caused me to worry because I have the elite form of the ACTN3 gene. But luckily elite athletes tend to live longer, with a twenty-nine percent increase in lifespan for swimmers and twenty-five percent for track athletes – translating to nearly five more years of life.[18]

Of course, there are always excuses for not exercising; if it isn't worrying about getting exercise-induced bronchoconstriction from overtraining it's the (misplaced) concern that breathing in all that filthy London air whilst exercising would be damaging. Back to Fuller, who told me that London air only crosses the threshold into unsafe, acutely damaging territory extremely rarely, putting paid to that one. He pointed me to some research on city rental bike schemes where they balanced all the risks – pollution,

---

* And presumably children's urine. The impact of other swimming pool detritus such as plasters remains unknown.

accidents, etc. – against the health benefits and found the benefits of the schemes outweighed the costs by a factor of fifty to one. Therefore, get on your bike!

SING WHILE YOU'RE WINNING

Not being a fan of Lycra, what else I can do? Some evidence suggests that singing improves lung function, but the studies tend to be small and poorly controlled.[19] It is hard to blind the participants to whether they form the crooning arm of a trial! I have, occasionally, been known to sing/boom* while in the car and one of the ideas I had whilst cooking up this book was to join a choir. But then, unsurprisingly, the time required to do the day job, support my children, speak to my wife from time to time and write a book mounted up and I gave up on the choir idea before conductor even tapped their baton, to the loss of all concerned.

A simple alternative, with the additional benefit of not inflicting my singing voice on anyone else, is to try breathing exercises. Some of the most commonly suggested ones include pursed lip breathing (breathing in the nose and slowly out through lips in a whistling position), diaphragmatic breathing (using your stomach not your chest to breath) and pranayama yoga breathing (timed breathing with a focus on exhalation). All of these offer an absolute mountain of instructional videos to watch. But as with singing, whilst many studies have been performed on the benefits of breathing, most lack robustness.[20] The results, when stacked together and taken with a pinch of science

---

* Opinions vary on the quality of this; between mine and everyone else's.

salt, slightly lean towards an improvement. The breathing exercise with the greatest reported benefit is probably pursed lip breathing, but most of these studies focus on people with severe respiratory diseases. Compared to the massive changes of not smoking or keeping pigeons, it is all extremely marginal. Ultimately, like cold water swimming, it is a 'you do you' situation: if you feel like you benefit from it, you probably will, but the science is sketchy.

Breathing is clearly vital to our life. And most of the time our lungs perform an admirable job to keep us alive. If we mistreat them, they eventually repay the insult by entering a cycle of self-destruction. Most of this destruction is driven by our ever-eager immune system picking a fight with the rest of the lungs. The lungs share similarities with the heart in this regard; initial damage plus inflammation triggers a cascade of damage that never resolves, slowly degrading the organ. Luckily, our lungs possess a degree of spare capacity; eventually the damage crosses a threshold where we notice a reduction in function. But now it is time to exhale that breath from the beginning of the chapter and move onto the next topic, cancer. We are not moving very far, as one of the most common cancers is found in the lung caused by the rubbish that we inhale.

CHAPTER 7

The Smoking Gun:
Killer Number 4 – Cancer

'A custome loathsome to the eye, hatefull to the Nose, harme-
full to the braine, and dangerous to the lungs.'

King James VI of Scotland, I of
England, 'A Counterblaste to Tobacco'

Of all diseases, we fear cancer the most. A survey conducted
in the US in 2000 reported that forty-eight percent of
respondents dreaded it more than any other illness; in the
same year, cancer placed as America's top terror – higher
than violent crime or nuclear attack. This has, however,
changed a bit in the last twenty years; Chapman University
in California runs an annual survey on American fears. For
the last seven years, corrupt government officials have
topped the list, beating loved ones getting ill, polluted
drinking water and even Russian nuclear missiles. Different
countries perceive risk in different ways – the number one
neurosis in an equivalent survey in the UK was . . . spiders.*

* We don't have any (native) venomous spiders in the UK, which
goes to show fear is irrational; likewise, the US is the twenty-fourth
least corrupt country out of 180 assessed (ahead of the UK, but
behind many, many others – it's no Somalia).

As with lung disease and stroke, death isn't necessarily the worst part of cancer. I'm not saying death is a good outcome, but the debilitation of terminal cancer and resultant loss in quality of life is what really concerns people.

Staggering numbers of people get cancer. In 2020, 288,753 new cancer diagnoses were made in the UK. The numbers work out such that every two minutes someone in the UK is diagnosed with cancer. The starkest statistic is that fifty percent of people in the UK will receive some form of cancer diagnosis in their lifetime. That's every other person; if you look up now from where you find yourself reading and count the people around you (including yourself) half of them will at some point get cancer.

This case load translates into a huge burden of death. When taken together, cancers account for one-quarter of all deaths in England in a typical year.[1] Broken down by type, the most common causes of cancer death are: lung, bowel, prostate (all men), breast (predominantly women, but some men), pancreas, cancer of unknown origin, oesophagus, liver, bladder, brain, lymphoma, leukaemia, kidney, stomach, head and neck, ovary, myeloma, uterus, mesothelioma and, rounding off the reaper's list, melanoma (skin cancer).[2] Globally, cancer accounted for ten million deaths in 2020.

This all makes for grim reading. Tragically, the fear of cancer contributes to the severity of cancer. When perceived as an untreatable disease, people are much less likely to get cancer tests because knowing makes it real. Many people prefer to be Schrödinger observing his theoretical cat, in the third unknowable state between life and death, and so delay diagnosis. Postponing testing worsens outcomes; early diagnosis saves lives. For example, almost all women diagnosed at the earliest stage of breast cancer survive for five years;

only three out of ten survive late diagnosis. Delaying testing inflicts a psychological impact too, caused by the stress of maybe or maybe not having a terminal illness. So, people end up carrying the nagging doubt. My experience with cancer reflects this cycle.

DAD'S SCRATCH

I had a mole on my neck that I had assiduously been ignoring for many years. I'd resisted comments from both my mother and my wife that I should get it checked, because I am a man and therefore an idiot when it comes to health.[*] Eventually, in 2011, following a direct instruction from a physician friend,[†] I went to have it looked at. Things progressed rapidly from there. I went to my actual GP, who immediately referred me to a dermatologist, who immediately referred me to surgery. All in about seven days. Before I knew it, I found myself in a surgical gown at St George's hospital, about two hundred metres away from where I had spent the last three years working, getting a local anaesthetic injected into my neck. The surgeon quickly removed the mole, and I returned home.

Just to up the ante, in an inexplicable decision that seemed sensible at the time, we had decided to pile as many life events into that same week as possible: we had moved house, hired a new childminder and started our son at primary school. The day of my operation a gruesome

---

[*] Not all men.
[†] Coincidentally, the same GP friend rushed my son to hospital in his first year of life when he was infected with RSV. One of my best health tips is *befriend a doctor*.

assault took place in the park behind our new house. I mention the assault only because apparently having a massive blood-soaked bandage on your neck when the police come to enquire about events is apparently a bit of a red flag. About two hours later a much more senior police officer dropped by to ask the same questions! Good way to impress the new neighbours.

Having reassured the local constabulary that I wasn't the Epsom ripper, I entered the period of the longest wait. Because after they removed the suspect mole the oncology team needed to check whether it was cancerous and if it had spread further. This was without doubt the worst part. In my experience, the unknown is far worse than the actual fact. The unknown is a massive great box of darkness without limit. It was a scary time for me and my family; sufficiently so that we had a family portrait taken 'just in case'. The pathology results came back and the mole was indeed cancerous but hadn't spread into the surrounding area. Though for safety, a wider section of skin needed removing. At this point, things got even more weird. While waiting for the drugs to kick in, I overheard the surgeon and the nurse talking about where they had met their other halves. The surgeon mentioned that she had met her husband at Cambridge, which pricked up my attention, because (and I am surprised that I haven't mentioned it yet[*]) I went to Cambridge too. I asked her which college and it turned out it to be the same as me, with us only separated by a year. The surgeon then stated, 'Oh yes, I remember; you look a lot older,' to which I didn't answer (on

---

[*] Surprised because going to Oxbridge, alongside being a vegan, driving an electric car and road cycling is something you are obliged to tell other people about within the first nanosecond of meeting them.

account of her having a very sharp knife next to my throat): 'cancer will do that to you.'

Because it removed a larger area, the excision didn't heal quite as neatly, leaving me with a somewhat jagged scar on my neck. Fortunately, thanks to the quick service of the NHS (and nagging of friends and family) I am in a position to write about this incident with just a pirate scar to show for it. From my experience I can share three pieces of advice: firstly, get checked often and promptly at any sign of danger; secondly, before getting checked, make sure your life insurance is in place as it is an absolute embuggerance to get insured after you show any sign of cancer; thirdly, as the man says, 'wear sunscreen' – we will come back to why later.

The fact that I am here to write about my cancer experience reflects one of the most important (and sometimes overlooked) facts about cancer: much of it is curable. Fifty percent of those diagnosed with cancer survive for ten or more years, with this survival rate more than doubling in the last forty years (up from twenty-four percent). This is a success story that should be shouted from the rafters, especially if it means more people seek earlier diagnosis.

WHAT IS CANCER?

As is traditional, let's start with the etymology: the word 'cancer' means 'crab' in Latin – you potentially knew this from astrology.* It received this name because the finger-like projections extending from a cancerous tumour as it invades the surrounding tissue reminded the omnipresent

---

* Confusingly, the 'ology is the made up one; astronomy is the science one.

Hippocrates of a crab; though he used the word 'carcinoma'. The word 'cancer' was first used by Celsus (a Roman doctor who lived four hundred years later) when translating Big H's Greek into Latin. The other term you may hear associated with cancer is oncology (an actual 'ology) that refers to the study of cancer: coined by Galen, who used the word *oncos* (Latin for 'mass') to describe tumours. Whilst naming rights sit with Hippocrates, he wasn't the first person to describe cancer. An ancient Egyptian scroll from 1600 BCE depicts cauterisation as a means to remove breast cancer; though I am a bit curious as to what the hieroglyphs for this would be – breast, crab, stick, mouth, fire, plier, blood?

One of the earliest cancer studies was made by Percivall Pott who found a link between soot and scrotal cancer in chimney sweeps. Another key individual was Rudolf Virchow, excellently known as the Pope of Medicine to his peers: presumably the pontiff, not the diminutive middle-order Surrey and England batter. Virchow was handy with a microscope and made a series of discoveries about cells, particularly during disease. A busy man, Virchow identified the mechanism of thromboembolism (blood clots) and how *Trichinella* parasites larvae grow in muscle cells; he even dabbled a bit in forensics. He also holds the remarkable record of being the only person to challenge someone to a duel where the weapon of choice was a sausage. In 1865, the prodigiously appetited Prussian Chancellor, Otto von Bismarck, had a beef with Virchow (in the Diet, but not about his diet*) and demanded satisfaction. Virchow got the

---

* If you ever seek the inverse of healthy eating, look no further than the Iron Chancellor, who had a legendary, prodigious appetite. Whilst campaigning in the Franco-Prussian war he ate mushroom omelettes, pheasants, turtle soup and a wild boar's head. Unsurprisingly, he

choice of weapons, offering Bismarck either a pork sausage or one infected with *Trichinella*. Not wanting to come off *wurst* for wear, Bismarck unsurprisingly backed out of the banger battle – no one likes bad sausage and Virchow was declared the *wiener*. Virchow plays a central role in the cancer story because he linked its origin to the misfunctioning of normal cells.

Fundamentally, cancer occurs when cells replicate out of control. All cancers start as a single cell – just one out of the possible forty trillion cells (4×10^{13}) that make up our bodies. You could say that this is extraordinarily unlucky. However, every day, two trillion of those forty trillion cells replicate – in that many dice rolls, we shouldn't be surprised that the occasional cell goes rogue. Raising the likelihood further, every time our cells replicate they must faithfully copy three billion (3×10^9) DNA letters without introducing mistakes, which because of the double helical nature of DNA actually means six billion letters per cell, all adding up to a staggering 2.4×10^{23} (nearly a septillion) new DNA letters a day. Unsurprisingly, this goes wrong, even without the undue stresses that we place on our cells. To avoid problems associated with faulty replication, nearly all organisms (except some viruses) have evolved proofreading processes, which check the newly copied nucleic acid for fidelity. This process is not infallible and over time errors will creep in. In other words, we shouldn't be surprised that half of us get cancer

---

often had digestive issues – to resolve these he reduced his food intake to a meagre diet of soup, a plump trout, roast veal and three large seagull eggs. He was also fond of the booze, drinking beer and wine by the gallon, sometimes in combination (Black Velvet, the Guinness and champagne cocktail, is sometimes called a Bismarck). And yet he lived till eighty-three!

at some point in our lives; what we should be surprised by is that, given the complexity of the system, the number isn't far higher.

ROGUE TROOPERS

Cancer starts when an individual cell breaks the shackles that prevent it from growing out of control, leading to uncontrolled and invasive replication. Cancer causes disease because our bodies contain a finite pool of resources and space which tumours disrupt. The human body is an incredibly complex living machine. It contains seventy-nine different organs, made up of two hundred different cell types. They all need to work together. If one fails, eventually all fail. Over time, individual cells in our organs need replacing due to deterioration or damage. In healthy tissue, (stem) cells replicate until they replace damaged regions and then they stop. The ability to stop replication is as important as the ability to start. Unregulated replication leads to out-of-control growth, forming the basis of a tumour, an abnormal mass of cells. If this tumour then starts taking up resources or space from other cells it can cause disease. Cancerous cells are not dissimilar to infections in this way; they parasitise the body, stealing from healthy tissue.

Shedding the shackles that constrain growth is the first step in the development of a dangerous tumour. But the real damage begins when cancerous cells escape from their original organ. Early on, tumours are constrained within the tissue in which they arise; at this stage they can be described as benign because they are not invasive. This can also be known as stage I. Survival rates for most cancers detected at stage I are high, for example ninety-eight percent of women will survive at least five years after a stage I breast cancer diagnosis. Stage II represents a transition, with the tumour expanding and beginning to spread out of control. Trouble really starts at stage III, when the tumour tips into being malign and invades other nearby tissues. One of the nasty tricks up malignant tumours' sleeves is to produce protein-eating enzymes (called proteases) that break down the boundaries between cells, creating *Lebensraum* for the tumour. Survival is still relatively high, because the tumour, though bigger, is at least a defined mass that can be removed or targeted: the rates of breast cancer survival for stage III are about seventy percent. In its fourth and final stage, the tumour escapes its original organ in a process called metastasis, meaning 'new place'. Metastatic tumour cells enter the blood or lymphatic system and move to other organs, seeding disease. Metastasis is bad news indeed: survival rates for stage IV breast cancer being less than twenty-five percent (The stages of cancer are shown in Figure 12).

What drives our genes to misfire and turn our cells into all-consuming invasive monsters?

CAUSE AND EFFECT

Nearly all diseases have both a genetic and an environmental risk. Cancer is no different; susceptibility can be inherited and faulty genes contribute to about five percent of cancers. Probably the most well-known inherited genetic association is between breast (and ovarian) cancer and the BRCA genes (often pronounced Brack-A), specifically BRCA1 and BRCA2. The gene name derives from BReast CAncer, because inheriting a singly faulty copy of either BRCA gene increases the risk of breast cancer in women

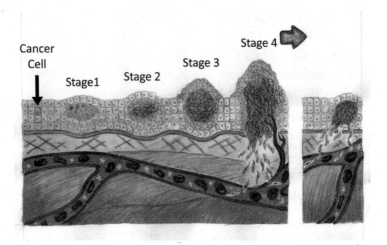

Figure 12. The stages of cancer. Tumours have four stages. In Stages 1 and 2, the cell is growing abnormally, but confined. In Stage 3 the tumour starts to invade local tissues. When it reaches Stage 4 the tumour becomes metastatic and spreads throughout the body.

about fivefold, from thirteen to seventy-two percent. In men, carrying a mutated BRCA2 also increases the risk of breast cancer (admittedly to a lower overall risk less than in women without the gene). I include this just to reiterate that it is possible to get breast cancer as a man, and that if your family has a history of it, it might be worth getting some genetic screening done.

On the flip side, environmental factors cause ninety-five percent of cancers. We will start by looking at the cancers considered preventable, which means they have a significant, avoidable causative factor. None of these risk factors should come as a surprise. And whilst much of cancer comes down to luck, good or bad, you can definitely stack the odds against yourself.

A LITTLE (DEATH) RAY OF SUNSHINE

In my case it was too much sun, a bit ironic given I live in the UK. However, my pallid skin is especially sensitive to solar damage. As well as all the light that we see and that plants use to grow, the sun produces waves across the whole electromagnetic spectrum from radio to gamma. Waves at the shorter, more energetic end of the spectrum – UV, X-ray and gamma – are all potentially menacing. Most of this dangerous light is absorbed by the ozone layer, which fortunately has begun to recover from the damage done by CFCs released from fridge freezers in the heady days of the 1970s.* Whilst there is still a bloody great big hole over the Antarctic – about 24.5 million km² (a thousand times the size of Wales) – the somewhat good news is that the hole has stabilised since its greatest extent in the year 2000 and may be even closing. Even without a hole in the ozone, some UV does get past the stratosphere; in the northern hemisphere UV levels peak around the summer solstice, so if you visit Stonehenge make sure you wear enough Woad to cover exposed skin. UV light encompasses a range of intensities with varying potential to cause damage. The intensity is proportional to the light's wavelength; UV can be split into lower-energy UVA (315–400 nm) and higher-energy UVB (280–315 nm). The highest-energy UVC never gets to the earth's surface but is the reason astronauts wear gold sun visors on spacewalks. All of this complex data is simplified to a single index in which a level of three and above requires

---

* A brief un-acknowledgement to the 'one-man environmental disaster' Thomas Midgely Jr – the inventor of both leaded gasoline and CFC.

additional protection; the UK only ever reaches eight on the scale, other places go higher, for example, Australia where two out of three people are diagnosed with melanoma before they reach seventy.

UV is dangerous because it can damage our genes. It mainly does this by fusing two thymine bases (the letter T in the genetic code) together into a single unit. This stops them from binding to the A (adenine) on the opposite strand, causing a nodule in the DNA backbone, affecting the cell's ability to read from it or make new copies. Most of the time these breaks are repaired by proteins specifically evolved to protect DNA from damage. Unfortunately, the cell sometimes botches the repair job, altering the DNA sequence and leading to a mutation. Since the order of the DNA letters affects the proteins being encoded, changes in genes can lead to changes in proteins. These changes can mean you make too much or too little of that protein. Importantly, whilst the word 'mutation' sounds scary, not all mutations immediately lead to cancer.

To put it in context, we need to consider the size and makeup of our DNA genome and the relatively rarity of a cancer mutation. Recall that our DNA isn't just made of genes stuck next to each other, like beads on a string, it contains coding regions (the genes) and non-coding regions (the rest). The coding part makes up only one percent of our genomes. Most of DNA mutations, if they do happen, happen outside the genes. Sometimes the changes do occur in our genes, but even that isn't immediately catastrophic. For most proteins in most cells, a small change doesn't really make a huge difference, for example if a single skin cell starts making more melanin, you aren't going to notice. However, changes in a subset of genes can cause cancerous problems. Fortunately, only 40 out of 45,000 human genes cause cancer when changed (the mutable ones are called oncogenes). The events leading to cancer at an individual cell level are rare. But, as mentioned, we have a lot of cells, and like the lottery, the more tickets you buy, the higher your chances of winning (or in this case losing).

To prevent cancer, the trick is to reduce the number of metaphorical cancer lottery tickets being bought. And for skin cancer the simple answer is to wear sunscreen. Sunscreen contains a complex mixture of chemicals grouped into two broad categories: physical and chemical blockers. Physical blockers will be familiar to fans of cricket as the white stripe of zinc cream daubed across Shane Warne's nose – they work by reflecting the UV rays. The chemical blockers act by absorbing the UV rays of the sun. You don't need to dredge the internet very deeply to find people worried about chemicals contained in sunscreen. However, these worries are mostly based on a misunderstanding (as usual) of toxicity studies. For example, one of the chemical blockers (oxybenzone) has been reported to interfere with hormone production. But (and this is one of those 'all caps' BUTs), this effect was seen only when rats were fed oxybenzone, and it would require 277 years of sunscreen use to reach the doses used.[3] Top tip – don't drink the sunscreen, however much it smells of coconut and summer holidays.

Our responses to the sun vary, and the most important variable is skin colour. People with darker skin have a lower risk of skin cancer because their skin contains more of the pigment melanin. Melanin absorbs UV light and releases it back as heat without damaging the DNA, protecting us. The amount of melanin can also change in response to the sun: tanning is the body's desperate response to make more of it in an attempt to prevent the damage from the UV. There is no such thing as healthy tanning, either from the sun or tanning beds. If anything, tanning beds are worse – indoor tanning can increase the risks of cancer by up to fifty percent (and that's before you pile on the other risks such as eye damage, skin ageing, burns and loss of consciousness[*]). It's also worth pointing out that a 'base tan' does not protect you against future burning: you are simply getting the DNA damage in early before exposing yourself to more UV later on. And yes, we do need some exposure to sun – because it triggers vitamin D production. But not very much – ten minutes in a T shirt at lunchtime will suffice (depending on skin colour) and anyway no one ever actually puts on the recommended amount of sun cream.[†] So, wear sunscreen.

UP IN SMOKE

'I know a man who gave up smoking, drinking, sex, and rich food. He was healthy right up to the day he killed himself.'

Johnny Carson

I admit the sunscreen section leant into the preachy. Sorry not sorry, I'm just getting started. Getting people to change behaviour is difficult. This becomes much harder when you are asking them to give up something addictive. Even when the risk is patently clear. I work at a hospital and I often watch people in hospital gowns, sometimes with drips

---

[*] There's even a risk of STI transmission. The virus HSV is spread skin to skin, so if not cleaned properly – sweat from the last Oompa-Loompa can infect you. Try explaining that one away, no darling I caught this in the tanning salon, *no* not that way.

[†] Just as no one ever eats a portion's worth of cereals.

attached, smoking outside in the freezing cold – and admire their commitment to getting lung cancer. And I am not without sympathy. But when it comes down to it, *not* smoking is an extraordinarily potent way to extend your life expectancy. We skirted around smoking as a cause of COPD in Chapter 6, but let's drill down into exactly *how bad* for you it is. Of the 3.3 million people across the world tobacco dispatched in 2017, it killed a third by lung cancer. And smoking is an equal opportunities carcinogen, it doesn't restrict itself to the lungs – it can cause fourteen other cancers, some in the airways (larynx, oesophagus, oral cavity, nasopharynx, pharynx) and others in remoter sites (bladder, pancreas, kidney, liver, stomach, bowel, cervix, blood and ovaries). Lighting up is the leading preventable cause of death in the UK. It causes more than twenty-eight percent of all cancer deaths in the UK and fifteen percent of all cancer cases. Then there are all the other 'side effects': smoking contributes to heart disease, stroke, COPD and erectile dysfunction (though admittedly that won't kill you).

Tobacco smoke contains a confounding concoction of carcinogenic chemicals. There are about seven thousand of them in every puff, many of which cause uncharacterised yet hideous harm. Some of the toxic chemicals come direct from the tobacco leaf itself, but others are added to make the cigarettes more palatable to the user. Which in and of itself is quite extraordinary: here is something palpably bad for you and sufficiently unpleasant that sweeteners need to be added to it – a spoonful of sugar makes the cancer go down as Mary Poppins probably didn't say. The list of potential additives is astounding, a scientific paper from 2004 listed 482, from the innocent-sounding such as banana extract and caramel to the forbidding and very chemically such as δ-Decalactone, 2,3-Diethylpyrazine and 2-Hydroxy-4-methylbenzaldehyde.[4] If you are worried about ultra-processed foods, you should definitely worry about the contents of cigarettes! When burnt, the resultant cacophony of compounds does tremendous damage.

The top three villains in this soup of evil are nicotine, carbon monoxide and tar. Nicotine is the addictive part of the smoke, ensuring people continue sparking up. Nicotine mimics the body's pleasure molecules such as dopamine and serotonin. Over time, it dampens the body's ability to signal through these pathways, which now needs more to hit the same response – hence the addiction. On the plus side, nicotine doesn't cause cancer, but it can raise blood pressure – contributing to strokes. The second compound, carbon monoxide (CO), gets produced when things are burnt without enough oxygen; it is the poisonous stepbrother of carbon dioxide (CO_2). It most commonly occurs with poorly installed gas boilers, cookers or fires, though legislation has dramatically reduced the fatal incidence of this in the UK to only forty deaths a year. Carbon monoxide has a grim history – the Nazi *Einsatzgruppen* used it in their mobile murder vans on the Eastern Front in 1941. Which raises the question: why would anyone voluntarily inhale it for pleasure? Poisoning following carbon monoxide inhalation occurs because CO preferentially binds to our blood's haemoglobin, essentially rendering our red blood cells useless. If too much is inhaled it suffocates you. Like nicotine, carbon monoxide doesn't cause cancer, but it does increase the workload on the heart, which is concurrently being damaged by nicotine.

In terms of cancer, the third component, tar, is the real problem. Tar is the burnt residue from the inhaled smoke.

'Reassuringly' it isn't the same as road surfacing tar, because who would inhale that? But cigarette tar is noxious stuff. An analytical breakdown of the individual components yields a list of chemicals each of which would need me to fill in a lengthy risk assessment if they were used at work. For example, tar contains formaldehyde, the chemical used to preserve dead bodies. In my lab we keep formaldehyde in a locked cupboard and only use it in a chemical hood that sucks away any trace. A bottle of formaldehyde has warnings that it can cause burns, cancer, organ damage and genetic defects. The rest of the list of tar chemicals is frankly too depressing to delve into further. Many of them can damage DNA (in the same way as the UV in sunlight), causing mutations and potentially setting lung cells on the path to destruction. Other chemicals in the tar kill airway cells, destroying the delicate structure of the lungs, contributing to COPD and generating an inflammatory environment that can fuel cancer. Smoking shares a horrible partnership with air pollution: both cause cancer and COPD.

The good news is that smoking rates have declined. British smoking peaked at eighty-two percent of adult men in 1948, probably as a consequence of the Second World War. Rates of smoking tripled during the war, no doubt helped by the free issue of cigarettes with the rations. Thinking about the long-term risk of smoking was probably quite low on Tommy's list of worries: the chance of getting wounded or killed in the poor bloody infantry was essentially 100 percent, so puffing on a Capstan was one of the few comforts available. Besides, very little about the long-term risk was known. Today, the smoking rate stands at 13.3%, or 6.6 million people in the UK. The decline in smoking results from a combination of education and legislation, with declines seen after the banning of cigarette adverts on TV (goodbye Joe Camel), increases in tax on tobacco products, health warnings on cigarette packs and smoking bans in public places. The rate of decline slowed in the 1990s when the introduction of new legislation also slowed. It will be interesting to see the next steps.

The government of New Zealand led the way in this regard. They aimed to be smoke-free as early as 2025. To achieve this goal, they introduced legislation that came into force in January 2023 that had a steadily rising smoking age, effectively meaning children born in 2009 would never be legally able to smoke. Except then they didn't follow through. A new government overturned the original plan. The UK undertook a consultation in autumn 2023 to introduce similar legislation. At the time of writing, this is still undergoing debate in the House.

One way to reduce smoking is through nicotine alternatives, such as gum and patches, that provide the addictive nicotine without all the harmful side effects of cigarette smoke. The most recent, and probably the most controversial, are vapes. A vape is a battery-powered device that creates an aerosol containing water vapour, nicotine, flavouring and other 'stuff'. Initially they looked like cigarettes, but designs diversified over time. There has been an extraordinary explosion of vape shops: I found three in the space of one hundred metres of one another in my hometown of Epsom. Vapes form an important part in smoking reduction.[5] [6] However, they are not totally benign. Vaping still carries the danger of nicotine and the risk could be greater because vapers vape more frequently than smokers smoke. There are also the dangers associated with regular inhalation of anything into the lungs. Finally, vapes may act as a gateway, targeted at children, with flavours such as raspberry, mixed berry, cherry and candy easily available. Vapes are still new, and the long-term epidemiology is uncertain. Overall opinion (at the time of writing) holds that they are net good because they replace cigarettes, but they present various problems, especially if children use them who would never have smoked. They feel a bit like swallowing a spider to catch a fly, and I suspect that if I wrote this in ten years' time, I would be describing smoking as a crazy thing of the past and focusing on the damage of vapes. For absence of doubt though, vapes are *much* better than smoking.

That is one vice down. Let's park the discussion about how alcohol, the other poison people commonly put into their bodies, also leads to cancer and return to it when we think more about the other damage it can cause to our livers and our lives.

YOU ARE WHAT YOU EAT

The final 'lifestyle' cancer is associated with diet. Predominantly, cancer is driven by obesity and the inflammation it causes. As should be clear by now, everything is interconnected; I will return to obesity in the chapter about diabetes.

But specific foods can also be carcinogenic. Processed meat and red meat top the list of bowel cancer-causing foods. Red meat contains haemoglobin (the protein that carries oxygen, which gives the meat its red colour) and this can cause cellular damage when broken down. Processed meat contains nitrites, added to prevent meat from spoiling; these are directly carcinogenic. As part of the collective madness induced by the boredom of lockdown, one of the diversions I undertook was making my own chorizo sausages.* They contained the double jeopardy of possibly getting botulism from insufficiently cured meat and giving you cancer; the sausages didn't even taste that good – recipe one contained so much fennel they basically tasted like meaty liquorice. Potentially they would have been good for duelling though. You can further increase the risk value of your sausage by grilling it. Grilled meat contains a family of compounds called heterocyclic amines which can damage DNA. Unsurprisingly, they are also found in cigarette smoke (because what isn't!). But surprisingly, if you marinate chicken before grilling it, you reduce the levels of these toxic compounds by about ninety percent; in a piece of surprising good news – healthy food can also be tasty.[7]

You might wonder, how does something as tasty as bacon cause me harm? A bit like my terrible punctuation, the problem lies in the colon. The colon (the penultimate tube before exit) isn't just a straight tube, it contains a crinkly surface, with many fleshy peaks and troughs. A bit like the lung the colon needs a very high surface area, increasing its ability to recover water from the digested food prior to defecation. This means that, whilst the tube itself is only 1.5 metres long and 5 centimetres in diameter, spread out it could cover half a tennis court. The colon is distensible, like a snake, and runs from your right hip, under the diaphragm, back down the other side, looping back to the rectum in the middle. In terms of cancer, the micro-anatomy of the colon

---

* Important note for all concerned: this is my book and I am pronouncing 'chorizo' with a hard English 'Z' sound and there is nothing you can do to stop me.

really matters. The crinkly surface comprises lots of little indentations called crypts. These U-shaped crypts grow from the bottom up; stem cells reside in the well of the U; they replicate and then move like an escalator to the tips of the U where they shed. The cells lining the colon undergo constant replacement. In fact, after blood cells, gut cells turn over the fastest.[8] You shed 10^{11} cells a day from your intestine – a large amount of what you poop is, in fact, you. The crypts spit out new cells to repair damage caused by the things you eat and the bacteria that inhabit your intestines. High cell turnover increases the risk of mistakes creeping into the DNA during copying. Aberrant replication in crypts can lead to outgrowths called polyps. As with other cancers, these can be mostly benign but occasionally tip into being more aggressive – colonoscopy can be used to find and remove these polyps. Food that irritates your guts (red, grilled and processed meat) increases the formation of the pre-cancer polyps, increasing the risk they will go rogue in the future.

What can we do? The obvious thing is to cut out red meat. The challenge (apart from bacon being extremely tasty) is the extremely low recommended amount. The NHS suggests seventy grams a day, and this isn't very much. Many people smash through this in a single meal, let alone a day: a cooked breakfast with bacon and sausages is 130 g, a portion of Sunday roast is 90 g. And seventy grams is the upper end of healthy; a study using data from the UK biobank says that cutting it to twenty-five grams a day (half a piece of bacon) reduces the risk of cancer by a fifth. No one is eating only half a piece of bacon for breakfast. The only realistic way is to remove meat from a substantial number of meals. As a British child of the 1980s, the idea of vegetarian food conjures a lingering grey and drab terror. But things have clearly developed, and one of the major changes I have made as result of researching and writing this book is to increase the number of vegetarian meals in the rotation. This includes three bean chillies, vegetable curries and tomato pasta, but I am open to menu suggestions!

Meat intake can in part be counterbalanced with fibre. We can increase fibre intake in the form of boring carbohydrates (brown pasta, rice and bread) or beans, vegetables and lentils. We should eat at least thirty grams of fibre a day. Fibre has a range of benefits (including feeding your pet bacteria, to which we will return), but it protects against cancer by moving everything along, giving the toxic meat less time in contact with your delicate guts. A lack of gut motility seems systemic across high-income countries, becoming so bad that the USA had a laxative shortage in 2023. Sadly, curries and spicy foods, while increasing velocity through the guts, are probably not protective – because the capsaicin in chillies is an irritant, damaging gut cells as it passes through at speed.

INFECTIOUS CANCER

The remaining fifty percent of cancers also have some form of environmental trigger (because everything does), but much of it is down to luck. Even with the most well-balanced diet and health lifestyle, you can get cancer. As Bill Hicks, the comedian and only person more pro-smoking than the American ad agencies of the 1950s said: 'Non-smokers die . . . every day.'

Of course, positive lifestyle changes (eat better, don't smoke, wear sunscreen, exercise, don't drink) reduce the risk of getting many cancers and other poor health outcomes, but they don't eradicate the risk. One major environmental risk for cancer is viral infection. Viruses must replicate in our cells to live. One of the body's defences is to self-destruct virally infected cells. But some viruses have found ways to inactivate this cellular suicide, leading to cells becoming immortal and therefore more likely to turn cancerous. Seven common viruses can cause cancer: EBV, HBV, HCV, HIV, HHV, HPV and HTLV. The first human cancer-causing virus isolated was EBV (Epstein Barr virus), discovered by Anthony Epstein, Yvonne Barr and Burt Achong; it remains unclear why Achong missed out on the naming rights. EBV is predominantly spread in saliva (it is sometimes called the kissing disease), and in the short term EBV causes glandular fever (mononucleosis). In the longer term it puts people at risk of leukaemia (blood cancer) by infecting and immortalising a type of white blood cell (also called leukocytes) – the B cell.

It may seem difficult to avoid viruses – they are pretty ubiquitous – but it is possible. The most striking intervention in the reduction of viral cancer is the vaccine for HPV (human papilloma virus). HPV causes warts (basically benign cell outgrowths) but can cause cancer, most commonly cervical cancer. The German virologist Harald zur Hausen made this link in 1976, winning him the Nobel Prize. Critically, he identified carcinogenic strains of HPV, leading to the development of vaccines. The impact of HPV vaccines has been remarkable: in the UK they have reduced cervical cancer in vaccinated girls by eighty-seven percent.[9]

SEX CANCERS

Several cancer-causing viruses are spread by sexual contact, but viruses are not the only risk factor for cancer in the reproductive organs. Alongside lung and bowel, the two most common types of cancer are breast and prostate. The initiation of either of these cancers is not well understood.

One reason for cancer to arise in the breast links to its function – milk production. The glandular tissue produces milk, and denser tissue makes cancer more likely, probably because it contains more cells that can turn bad. Some risk factors are unavoidable, such as age, ethnicity and inherited genes; others are potentially avoidable, such as obesity, alcohol intake, contraceptive pills and HRT. But each risk is small individually and it is more about total risk of all these things combined than any one thing (unlike smoking and lung cancer). The main way to avoid the danger from breast cancer is screening and monitoring. Catching lumps early before they develop into full blown cancers significantly reduces the risk of death from breast cancer. This can be through self-assessment, but also through screening programmes and mammograms (offered by the NHS). Early detection is associated with a forty-one percent reduction in death.

A similar state of affairs applies to the prostate. At a simplistic level, this funny walnut-shaped lump encircling the male urethra performs a similar function to the breast: it produces fluid that is excreted from the body. Obviously, the purpose of the fluids is different, but the risk factors for cancer are not dissimilar: age, ethnicity, family history, some genes (including BRCA1/2 like breast cancer), obesity and diet. However, unlike breast cancer prostate cancer lacks a straightforward screening programme. The most common screening method is measuring PSA (prostate specific antigen); but it's not great. The PSA test is good at finding positive cases (highly sensitive) insofar as most people with prostate cancer will have elevated PSA (though one in seven people with prostate cancer don't have elevated levels); however, it is likely to give false positives (poor specificity) because all men have PSA in their blood, and other causes such as a urinary infection can increase it. It's not to say it isn't worth testing, but be aware the results are not definitive.

Prostate cancer and breast cancer can both be treated by surgical removal. But surgery can significantly reduce quality of life. However, I have discovered some good news (for prostate cancer at least). And having learnt my lesson from writing about risqué things in my first book, I am going to put a deliberate page break here, to ensure my family don't accidentally read the next part.

Those of a sensitive disposition should look away now.

Ejaculation. That seems to be one way to reduce the risk of prostate cancer. Researchers at Harvard set up a long-term study of men's health called the Health Professionals Follow-Up Study in 1986, with a range of questions then correlated against health outcomes. The study followed 31,925 men. One of the questions related to ejaculation frequency (leading to the study's other, better name – the Harvard Ejaculation study). The results indicated a clear correlation: men who ejaculated more than twenty-one times a month reduced their risk of prostate cancer by two-thirds compared to those who only ejaculated seven times a month.[10] Simply put, abstinence is bad. According to the esteemed reference work *Roger's Profanisaurus* (subtitle *War and Piss*)[*], there is such a thing as a dangerous sperm build-up (DSB).[11] Though it's worth considering the dangers of progressive outer retinal necrosis (whose acronym is exactly what you think it might be).[†]

---

[*] Which I had hidden in plain sight on our bookshelves during the COVID pandemic, meaning every TV interview done by my wife about the value of vaccines had subliminal messaging sneaked in.

[†] And I'm not making it up, PORN has its own ICD code – B02.3.

Right, Mum, Dad and the in-laws, you're back in the game.

There is one other risk that sits across all cancers, and that is inflammation. As with other chronic conditions, general systemic inflammation makes things worse. The first person to identify a link between the two was the frankfurter-fighting Rudolf Virchow in 1863. All sorts of things can drive cancerous inflammation: stress, grilled meat, bacterial infection (particularly *H. pylori*, the cause of stomach ulcers), a 'bad' microbiome, environmental pollutants, night shift work, obesity. Inflammation is a complex process, and it can assist tumour formation in a number of ways, most of which are only beginning to be unpicked by researchers. Simple anti-inflammatory drugs such as aspirin can reduce the incidence and mortality associated with cancer.[12]

CURE FOR CANCER

But clearly, just taking a couple of aspirin, whilst surprisingly effective for many things, isn't going to cut it for most cancers. It should be recognised that, like the common cold, cancer isn't just one thing: it is a range of diseases across different cell types, with different causes, mechanisms and outcomes. The question isn't whether there is a cure for cancer, but a cure for *my* cancer. Increasingly, the answer is yes.

The earliest approach was surgery, cutting out the offending mass. Prior to the discovery of anaesthetics and the understanding of the need for hygiene, cancer surgery was brutal and chances of survival were slim. The development of germ theory by Pasteur and Koch and the application of it by Lister to develop basic sterile technique increased the likelihood of surviving surgery (two out of five patients died before 1865). Surgery has become increasingly refined as the ability to image tumours in situ improves and less invasive techniques are adopted.

Surgery was used (badly) for a whole range of diseases in the nineteenth century: the first cancer-specific therapy used radiation, called radiotherapy. This built on the findings of Maria Skłodowska-Curie (more commonly known as Marie Curie), the remarkable Polish/French physicist who won two Nobel Prizes. Sadly, Curie's science killed her – the radiation she discovered slowly depleted her red blood cells. Within a very few years of Curie's breakthroughs, radiation was being used to attack skin cancers. Radiotherapy works by completely shredding the DNA in the cancer cell, causing it to die. In its early form radiotherapy was a pretty blunt instrument. Henrietta Lacks – after whom the incredibly important HeLa experimental cell line is named – was treated by having two radium tubes inserted where her cancer was; these would have indiscriminately fired radiation through her tissues, causing lots of collateral damage. Modern approaches are much more focused: instead of a single high dose-beam, oncologists fire several lower-dose beams at the cancerous mass from different directions, concentrating the power on the tumour (a bit like crossing the beams in *Ghostbusters*). Radiotherapy causes side effects, such as sore skin and tiredness, associated with the damage done by the radiation.

Another mainstay of current cancer therapy is chemotherapy. Surprisingly, the roots of chemotherapy lie in chemical warfare. In the Second World War, in spite of being the 'good guys', the US Army developed a more potent mustard gas. This horrible stuff got its name from its smell, which is a bit like mustard, but that's where the similarity ends; it is carcinogenic, mutagenic, blistering and often fatal. Mustard gas made Wilfred Owen question the sweetness of dying for one's country.

> If you could hear, at every jolt, the blood
> Come gargling from the froth-corrupted lungs,
> Obscene as cancer, bitter as the cud
> Of vile, incurable sores on innocent tongues
>
> Wilfred Owen, 'Dulce et Decorum Est'

The need for the US to 'augment' mustard gas remains unclear, as it's not as if the original lacked potency – the use of mustard gas in the Iran–Iraq war caused 100,000 casualties. They named the 'new and improved' compound nitrogen mustard. In 1943, a bombing raid caused an accidental release of mustard gas in the Italian port of Bari. The Allies stored the gas as a just-in-case retaliatory measure lest the Germans used it first. In what is sometimes described as a 'lucky accident',* it was discovered that mustard gas reduced the number of white blood cells. This was identified by a US Army doctor, Stewart Alexander, who had the foresight to collect samples. Alexander's observation led to the use of the newer nitrogen mustard to treat lymphoma (white blood cell cancer).

Researchers have developed a number of different chemo-drugs, all with the aim of killing the cancer cells but not other healthy cells. This is tricky, because unlike an antibiotic which can target unique bits of bacterial biology not

---

* Admittedly not that lucky for the 83 people who died or 628 people who suffered chemical burns.

present in human cells, cancer cells *are* human cells, using the same biology. One thing that differentiates cancerous cells from most healthy ones is the speed at which they multiply, so chemotherapy drugs often target rapidly dividing cells. But this leads to many of the side effects, as certain other cells in your body also replicate quickly, such as hair cells, immune cells and the gut lining. Two of the most common side effects of chemo, hair loss and susceptibility to infection, are both caused by these off-target effects. Chemotherapy has evolved since the use of a chemical warfare agent, becoming more targeted and less toxic, but the real game-changer has been to harness the power of the immune system to fight cancer.

SELF-DEFENCE LESSONS

So far, when thinking about non-communicable diseases, misguided immunity has been part of the problem. And to some extent that is true in cancer – inflammation is definitely a risk factor for the initiation of tumours and the acceleration of their growth. However, once a tumour arises, the immune system is needed to fight it, and many tumours never get going because the immune system detects them and nips them in the bud. But tumours are immunosuppressive – shutting down our ability to identify and eradicate them. Turning the immune system back on in a controlled and focused way through immunotherapy has led to huge breakthroughs, and we are only at the beginning of the modern immunotherapy era.

An American surgeon, William Coley, made the earliest steps to harness the immune system as a cure for cancer by injecting bacteria directly into tumours. The first patient was an Italian migrant and drug addict called Zola, and following bacterial injection in 1891, his tumour dramatically vanished. Building on this Coley developed his toxins™, but sadly, his subsequent trials had a twenty percent mortality rate – injecting live bacteria into very sick people will have that effect! Coley suffered a bit of a PR failure and his approach was rejected in favour of radiotherapy, the wonder of the age.

A century later, researchers revisited Coley's work and eventually developed a more refined approach: immunotherapy. It built on the work of James Allison, at the University of California in Berkeley, on the regulation of T cells. He discovered a molecule on the surface of T cells called CTLA-4 (cytotoxic T-lymphocyte-associated protein 4) that can deactivate them. He linked the idea of T cell deactivation to cancer cell immune suppression by demonstrating that blocking CTLA-4 increased T cell killing of tumour cells. In short, increasing immunity cleared the cancer. In parallel, another researcher, Tasuku Honjo, working at Kyoto University, discovered a separate molecule that also shut down a T cell response called PD-1 (programmed cell death protein 1). Since both of these molecules turn off the immune system, blocking them should restart it – a bit like a minus and a minus makes a plus (or a double negative for those who don't do maths). Antibody drugs, called immune checkpoint inhibitors, have been developed to target both CTLA-4 (ipilimumab) and PD-1 (nivolumab). They have had an extraordinary impact on solid cancer treatment and led to Nobel Prizes for Honjo and Allison in 2018.

Allison and Honjo's checkpoint inhibitors work by boosting the existing immune response. An alternative is to retrain the immune system to recognise the cancer using therapeutic vaccines. These subtly differ from the vaccines you routinely get to prevent infections. This is because infectious disease vaccines are given *before* you get the pathogen, therapeutic vaccines are given *after* you develop the disease, in this case cancer. There is a huge amount of ongoing research in this area, including the use of RNA, the same platform that got (most of) us out of the COVID pandemic scrape. Hadi Sallah and Rob Cunliffe in my lab investigate ways to make RNA vaccines more effective. Whilst no cancer vaccines are available (yet), an even more futuristic approach has been developed – generate artificial T cells that specially attack the cancer. These CAR-T Cells are T cells biologically engineered from a patient's own blood and trained to attack that person's cancer; an alumna from my team, Katie Flight, now works on them. Data suggests that CAR-T provides a thirty to forty percent success rate – but one round of treatment sets you back about $300,000.

Price is one of the biggest problems with the new cancer therapies. Even older treatments such as chemotherapy (about £3,000 per course) cost a substantial amount of money.[13] The UK NHS spends about £1.5 billion on cancer annually – three percent of its total budget.[14] The costs of newer drugs can make the eyes water. Data from the US from 2018 estimated that antibody treatments cost nearly $100,000 a year – three times the average American annual salary .[15] This then stacks on the disproportionate incidence of cancer in areas with higher deprivation in a horrific feedback loop. The American states with the highest poverty also have the highest cancer death rates. Poverty is also a major contributor to cancer in the UK; for now, at least, we are fortunate in having a national health service that does cover most of the expense,[*] but in a cash-strapped system (with ever increasing expenses), tricky decisions need making. This falls on the shoulders of NICE (National Institute for Health and Care Excellence). It has to choose what can and cannot be used by the NHS to treat people. Decisions are predominantly based on value for money, which in its simplest form asks, 'How many lives will drug X save vs how much it costs?' A treatment that saves one person but costs £300,000 will get a lower priority than one that saves one hundred people but costs £3,000. This 'for the greater good' model serves society best but can seem heartless on an individual level.

In the end, prevention is nearly always better than cure, and certainly cheaper. The list of 'things that can cause cancer' is long and oftentimes ridiculous. A bit of ropey epidemiology and a catchy headline often flags things in the public eye without their really having a causative effect. Some of the things reported to give you cancer include: blowjobs, bras, broken hearts (not clear if those three are related), both having children and being childless, crayons, being left-handed, Worcestershire sauce (the impact of Sheffield's superior Henderson's relish remains unrecorded), being a man and being a woman. The WHO lists 1,108 potential cancer-causing agents; it categorises them from the most serious 'carcinogenic to humans' (Group 1), through 'probably carcinogenic to humans' (Group 2A), to the least serious 'possibly carcinogenic to humans' (Group

* Some of the treatment-associated costs fall on the patient – particularly travel, but also parking, which is inexplicably not free for patients or staff in UK hospitals (maybe not that inexplicably – NHS England made £280 million from parking in 2018/19, which is quite a lot of chemo doses).

2B). Bear in mind that these groups don't necessarily reflect the dose required – plutonium and bacon both belong to Group 1, but clearly you need to eat a lot more bacon than plutonium to get gut cancer. One of the more avoidable carcinogens is space travel, unless you are one of the 644 people[*] fortunate enough to have left earth's orbit.[†] The breadth and length of the list of 'causes of cancer' and the incidence of cancer (remember one in two of us will get it at some time) to me says that avoiding some of the more niche causes isn't time well spent, or even possible (for example, it is quite tricky biologically to be neither a man nor a woman).

As with all the diseases of ageing, prioritise the big four – eat less, do more, don't drink, don't smoke. Again, we see the layering of the same risk factors. But my quest for new life-extending hacks has slightly stalled – thus far I have added wear sunscreen, don't keep pigeons and ejaculate more. Maybe I will find solutions in other organs.

Much of the fear that cancer engenders comes from the long-term disease it causes and the devastating changes to people's lives. Brain cancer tops the list of people's fears because of the intimate interweaving of our brains with our very being. Tumours in the brain present a huge challenge because of the complexity of the tissue in which they reside, and we will turn to this most remarkable of organs next.

---

[*] The number varies on how you define space: the FAI uses the Kármán line 100 km above sea level (644 people); the USAF uses 50 km (681 people). Both Virgin Galactic and Blue Origin cross the Kármán line if you can afford the ticket.
[†] I like the irony of billionaire brogrammers injecting themselves with eternal life serum and then flying into space to get cancer.

CHAPTER 8

A Brain Is Only as Strong as Its Weakest Link: Killer Number 5 – Dementia

'We are not interested in the fact that the brain has the consistency of cold porridge.'

Alan Turing

We have looked at blood, lungs and the heart, which for unknown reasons is considered the seat of the soul, when it is actually just a really good muscle. What really makes you you – more than anything else – is your brain. This explains why diseases that affect it prove so devastating, both to the person and their loved ones. To compensate for the bleak start to this chapter, let's move to something slightly more fluffy – kittens. Who doesn't love kittens? I want you to picture a kitten in your mind. You should now have an image of a lovely fluffy kitten in your head – except for the three percent of people with a condition called aphantasia.* Imagining fluffy kittens is just one of the remarkable abilities of the three-pound lump of protein and fat that resides inside your noggin; even more remarkable when you

---

* This is not the same as synaesthesia, where people experience multiple sensory pathways simultaneously, for example seeing colours when listening to music.

consider it is nearly seventy-five percent water. So maybe the homeopaths are right, water can have memory too?* Even when not imagining fluffy kittens, our brains are pretty busy, controlling bodily functions, contemplating the universe and doing everything in between. There is a pretty good reason brain death is predominantly used as a definition of death – without the brain, we cease to exist.

The brain is superbly complex. I am only going to give the shallowest of overviews of how it works here. While we have a pretty good understanding of how the heart and the lungs work, the innermost workings of the brain still lie beyond our grasp. I love the thwarted circularity of how we can use our brain to think about how our brain thinks without really understanding how it thinks. The brain contains two types of cell: neurons and glia. The neurons are often portrayed as the stars of the show; our heads contain about 68 billion of them (Figure 12). They look like an uprooted tree with three parts, the cell body (the leafy foliage), the axon (the tree trunk) and the axon terminal (the roots). The 'foliage' of the neuron is made up of dendrites, little cellular arms through which they contact other neurons. The 'trunk' of the neuron carries signals from one cell to another; it is coated in an insulating layer called myelin, (remember this – it plays a key role in MS). The neuron transmits information to other cells further down the line through the 'roots' or axon terminals.

Individually, neurons aren't that special. But through their connectivity they sparkle. The vast number of connections (hundreds of trillions) make thought, movement, breathing and anything else we choose to do possible. Brain

* They aren't.

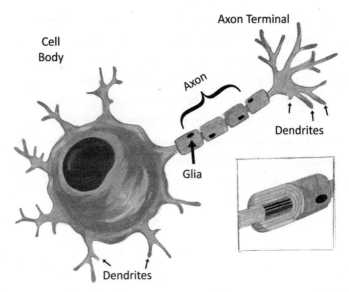

Figure 13. Brain cells. Shows a neuron and the three parts – cell body, axon and axon terminal as well as the dendrites that talk to other cells. Cutaway shows the supporting glial cells and the sheath of myelin.

cells work through a combination of electricity and chemistry. Signals track along neurons from the dendrites to the axon terminals as electric currents (called action potentials). When the electric signal reaches the terminal, it triggers the release of chemicals called neurotransmitters. Our brains contain over one hundred of these chemical messengers. They subtly modify what happens to the next nerve: some activate, and some dampen; acetylcholine, glutamate, GABA, serotonin and dopamine are the five most important. They all play a vital role in brain activity, and, not surprisingly, narcotics affect their function – heroin and cocaine affect dopamine (pleasure and reward); alcohol affects serotonin (impulses) and marijuana affects acetylcholine (motivation). Glutamate accumulates when we think, which is why we feel tired when concentrating – a bit like lactic acid build-up in the muscles.[1]

The glia often get overlooked as the neurons' frumpy relatives – but they are equally critical to brain health. The word 'glia' comes from the Latin for 'glue', because they were originally believed to serve a single, solitary function – holding neurons in place. Whilst one type of glial cells, oligodendrocytes, does play a structural role, wrapping around neurons making the insulating myelin sheath, other glial cells play a much more sophisticated role in brain health. For example, the motile microglia act as the brain's immune cells, eating infected or damaged neurons. The astrocyte, another glial cell, also performs a housekeeping role, hoovering up excess neurotransmitters and tidying neural synapses. Both microglia and astrocytes need restraining on a tight leash; when inflammation overstimulates them, they contribute to many long-term brain diseases.

BRAIN AGEING, REVISITED

Whole libraries of books have been written about what the brain does and how it does it (Figure 14); our story focuses on how the brain changes with ageing and disease. At birth, the brain starts at about a quarter of its final size, though it undergoes such rapid expansion that by the age of three it has grown to eighty percent of its adult size. It is nearly full-grown by age five, explaining toddlers' comically large heads relative to the rest of their bodies. Brain growth comes not from having more neurons but from their making better connections (increasing in complexity not number). Initial brain development focuses on learning basic motor control – walking, smearing Weetabix into the carpet, smiling in a sufficiently cute way that they aren't abandoned at the nearest service station. In the next developmental stage, we learn how to file thoughts – 'the obsession with dinosaurs stage'. Curiously, the next important step actually involves a pruning of paths. Adolescent brains are particularly plastic, reinforcing some pathways at the expense of others, shaping the future personality (which explains much stereotypical teenage behaviour: moodiness, tiredness, but also experimentation).* At the far end of this stage (early twenties) the brain contains nearly 500 trillion connections, a near infinite cat's cradle of interdigitated dendrites. We stay in this relatively happy state till

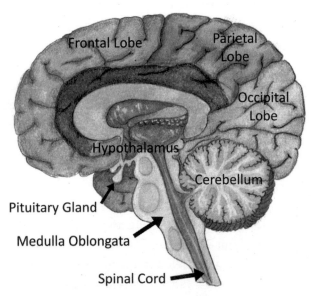

Figure 14. The brain. The seat of ourselves, a wonderfully complex mass of connections and chemicals that makes us, us.

---

* *Inventing Ourselves* by Professor Sarah Jane Blakemore explains this much more lucidly and is a great scientific primer about teenagers and why their brains aren't actually misfiring.

sometime in our late forties when our brains begin to shrink. Total brain volume dwindles with age, with the biggest decreases in the regions associated with problem solving and memory (the frontal lobes and the hippocampus).

HERE WE ARE NOW

What do these changes mean in healthy ageing? Reflecting every other system of the body, as we age our brains work less well, particularly in cognitive skills that require the manipulation of information, reasoning and the acquisition of new knowledge. Hence, I can recall the lyrics of the *Nevermind* album by Nirvana, first heard in 1991 when I was full of teen spirit, but fail to remember why I went downstairs thirty seconds ago.

Cognitive skills can be split into two types – crystallised abilities (ones gained in the past) and fluid abilities (those that need new thinking to solve). Fluid ability can be inferred from the speed with which you can solve problems using new information – like an escape room (or *The Krypton Factor* depending on your age). On average this declines in a linear fashion from our twenties onwards, leading G. H. Hardy to describe maths as 'a young man's game' in *A Mathematician's Apology*,[2] stating that he could not think of 'an instance of a major mathematical advance initiated by a man past fifty'. Whilst cognitive abilities decline, crystallised abilities, particularly vocabulary, actually increase well into the sixties on average. Which helps to explain why I can crush my teenage son at Boggle but lose to him at chess.

It was originally assumed that neuronal loss caused declining mental acuity. But this is only true up to a point;

newer imaging techniques show that the cells themselves aren't lost, rather the connections between them break. Like childhood friends, our neurons slowly drift apart. There is a pretty consistent downward trajectory of connectivity, and people are considered to have dementia when synapse loss exceeds forty percent. Fortunately, the normal rate of decline is pretty slow – about 0.25% per year. Assuming peak connectivity of ninety percent when aged 25, we should only cross this threshold aged 130 – by which time any number of other vital organs will probably have packed in.[3] Problems arise when the rate of decline accelerates (Figure 15).

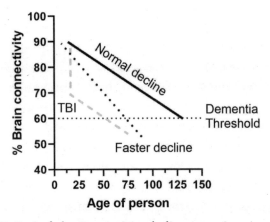

Figure 15. Brain fade. Connectivity declines over time; in traumatic brain injury (TBI), there is a dramatic drop in connectivity meaning the threshold for dementia is crossed at a younger age. Likewise conditions that accelerate decline lead to earlier onset dementia. Graph and concepts based upon data and hypothesis presented by R. D. Terry and R. Katzman.

VILE BODIES IN THE BRAIN

Dementia is a cluster of different symptoms, most commonly: memory loss, reduced speed of thought, failure to use language, reduced understanding and problems performing daily activities. As an umbrella term, dementia covers many different subtypes of the same condition. The predominant form of dementia in the UK is Alzheimer's disease, which contributes to nearly sixty percent of cases – and the terms 'Alzheimer's' and 'dementia' are often (incorrectly) used interchangeably. The other leading causes include vascular dementia (twenty percent), dementia with Lewy bodies (fifteen percent) and frontotemporal dementia (two percent). In the UK in 2019 there were approximately 850,000 people living with dementia, the majority of whom are elderly. Depending on your brain half full, brain half empty perspective, the proportion of over-sixty-fives with dementia is quite high (nearly one in fourteen) or quite low (thirteen out of fourteen don't have it). There is predicted to be a slow but steady rise to around 1.5 million people living with dementia in the UK by 2040. Worldwide, 36 million people live with dementia, much of which goes undiagnosed. The cost of dementia in the UK is £25 billion, the costs split two-fifths to the state and three-fifths to the families, a large component of which comes in the form of unpaid care. As a progressively debilitating disease, dementia is a major cause of death. Memory loss may be the first sign of its onset, but synapse loss affects all brain functions. Dementia can contribute to fatal secondary complications such as falls or infections, but with proper care these can be avoidable. Dementia directly kills you when the neurodegeneration damages the medulla (the part of the brain that controls involuntary functions such as heart rate and breathing); if this is damaged, the body shuts down and dies.

Connectivity is key, but how do the different types of dementia cause its loss? Whilst it isn't a major cause, we will start with a special mention for sport-associated dementia, as one of the more avoidable forms. Repeated blows to the head can cause a condition called chronic traumatic encephalopathy, a neurodegenerative disease that leads to dementia. Once known as *dementia pugilistica*, or simply as being 'punch-drunk', it has long been recognised as an occupational hazard of boxing. However, more recently it has been diagnosed in footballers (both association and American flavours) and rugby players including Steve Thompson, hooker in the World Cup-winning England team of 2003.[4] The acceptable level of head injury needs debate – where should the line be drawn between letting people do the thing they want (and which could earn them a living) and protecting them from the damage it does them long term? In the case of recreational drugs, that line has been drawn fairly clearly, but what should be done about contact sport?

Fundamentally, the brain is quite fragile and knocking it about disrupts the connections: traumatic brain injury increases the likelihood of dementia later in life two to four times. Traumatic brain injury features highly in the long catalogue of grim after-effects of the post-9/11 wars in Afghanistan and Iraq. Thanks to improved body and vehicle armour, far fewer people died from roadside bombs, which is the good news, but being violently tossed around an armoured car leaves a lasting imprint on the brain. In the cheerily titled 'Modern Warfare Destroys Brains' Colonels Warren Stewart and Kevin Trujillo describe traumatic brain injury as the 'signature injury of the war on terror', with twenty percent of US veterans having some form of brain injury – amounting to nearly 200,000 military personnel since 2000 (twice as many as are in the entire British Army).[5]

Whilst tragic, traumatic head injury is only a minor player; vascular dementia is far more frequent. Vascular dementia can be likened to a slow, ongoing stroke; strokes are caused by a massive failure of blood flow to the brain, while vascular dementia is a slow decline due to smaller vessels failing. The symptoms depend a bit upon which parts of the brain the blood fails to reach. Different regions of the brain control different aspects of our lives, for example the front does smell, the back does vision. Vascular damage is the sole cause of dementia in about five to ten percent of cases, but most of the time it contributes to other types of dementia, in particular associating with the types of damage seen in Alzheimer's disease.

FRAU DETER'S FORGETFULNESS

Alzheimer's disease is the most common of all the causes of dementia. It was first characterised by Dr Alois Alzheimer, a German psychiatrist who worked in the Frankfurt Institution for the Mentally Ill, colloquially known as the *Irrenschloss* or 'Castle of the Insane'. In 1901 he began recording the case of Frau Auguste Deter, admitted following a rapid decline in memory, development of paranoia and insomnia. Her husband Carl, a railway worker, couldn't afford her care and Alzheimer offered to carry on treating Auguste for free if he could examine her brain after death. She died in 1906, and Alzheimer cashed in his deal with Carl; on dissecting her brain Alzheimer found plaques – clusters of protein – and tangles of fibres which he associated with the onset of dementia. Remarkably, Auguste's brain (or at least slices of it) are still preserved, and in the 2010s, German neuroscientists discovered mutations in her DNA associated with the risk of Alzheimer's disease.[6] Alzheimer revealed his findings to little fanfare at a meeting of Southwest German Psychiatrists in 1906, finding himself rapidly ushered off the stage at the end of his talk, in favour of a speaker addressing compulsive masturbation. As with many scientific breakthroughs, a different, less well-known researcher undertook similar work in parallel. In 1907, Czech psychiatrist Oskar Fischer reported findings of plaques in sixteen post-mortem brains. Tragically the SS murdered Fischer in the Theresienstadt Ghetto in 1942 and his contributions were lost because of nationalism and anti-Semitism.

The plaques that Fischer and Alzheimer observed contain a tangle of proteins. The main culprit is amyloid beta (Aβ), a shortish protein fragment of the Amyloid-beta precursor protein (APP). Whilst the APP protein is naturally present in the brain, its breakdown products cause problems. Virchow (the sausage-wielding duellist and tumour pioneer) also observed these plaques, but he didn't link them with dementia. John Hardy and David Allsop made the connection between Aβ and Alzheimer's disease at St Mary's hospital – a place famous for Sir Alexander Fleming's discovery of penicillin, where I spent eighteen (long but mostly happy) years of my life doing research.

Amyloid beta (Aβ) is a core biological hallmark of Alzheimer's disease, but the exact mechanism by which it causes the disease remains to be ascertained.[7] In Alzheimer's Aβ accumulates faster than it can be cleared. Accumulation starts in the outer areas of the brain and slowly creeps inwards. Blood supply to the brain helps remove the Aβ, which may explain the overlap between vascular dementia and Alzheimer's. If the blood supply fails, it can't flush away the brain's waste. Immune cells break down Aβ: housekeeping astrocytes chew it up, but as they age, they get less efficient at digesting their amyloid meal. Aβ can also lead to inflammation, overstimulating the microglia and causing more damage (Figure 16). Alzheimer's shares parallels with other diseases such as COPD: ageing, chronic inflammation and stress reduce the ability of macrophages to eat toxic material and this in turn reduces the function of the organ. The brains of people with Alzheimer's disease show marked neuroinflammation (inflammation of the nervous system). Inflammation following infection may also contribute to dementia. Whilst highly active antiretroviral drugs mean that HIV infection doesn't inevitably lead to AIDS, HIV can cause dementia or at least some form of cognitive impairment. This is probably because the virus enters the brain, causing damage. Dr Merle Henderson in my lab is doing a PhD investigating the links between HIV, inflammation and cognitive impairment. One of the major shifts in last forty years is that (in high income countries at least) HIV has moved from a lethal infection to a comorbidity.

The causal role of inflammation in Alzheimer's needs further investigation but the accumulation of Aβ is the first event in the destructive pathway, leading to the recruitment of other sticky proteins such as tau, which then causes protein tangling, damaging the interconnections between neurons and directly killing brain cells. In summary, too much Aβ is bad, setting off a dangerous cascade of other sticky proteins, leading to holes in the brain and loss of cognitive function.

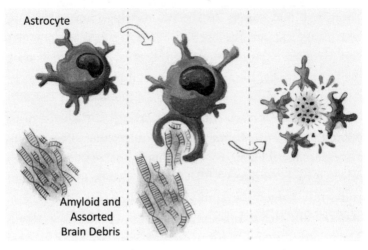

Figure 16. Just a wafer-thin plaque. Brain macrophages (astrocytes) eat up the amyloid till they are full and eventually explode, causing further damage.

For the last 120 years, there had been little in the way of therapy for Alzheimer's disease, but then suddenly, like the famous London buses, three treatments arrived at once. The breakthrough comes back to the fundamental research that identified Aβ as the primary trigger for Alzheimer's. Removing the build-up of toxic Aβ plaques can reduce the damage to the brain and slow the onset of dementia. The approach is a bit like vaccination and cancer therapies, harnessing the power of the immune system to protect us from disease. In this case, the drugs bind to Aβ and help the immune cells in the brain break it up before damage occurs. All three drugs are antibodies and hence have ridiculous names (donanemab, lecanemab and remternetug), which some might consider unkind given they are for people who struggle to remember things. Data from Eli Lilley reported that donanemab could reduce the decline in the ability to carry out daily tasks and reduce disease progression by twenty to thirty-five percent.[8]

COGNITIVE DECLINE AND FALL

Asides from dementia, there are other debilitating brain diseases associated with ageing including Parkinson's Disease, MS and epilepsy. One of the most unpleasant ones is Huntington's disease. This can occur as early as your thirties; if I were to have it, I probably would know by now. Huntington's disease was historically called Saint Vitus' Dance.* George Huntington, an American physician, first described it in a research paper at the age of twenty-two, observing it whilst accompanying his father and grandfather, also doctors. Based on the cohort of doctors I know, inheritance of a medical career is more dominant than brown eyes or tongue rolling. But remarkably, doctor is not the most dynastic job – nepo babies are found in a whole range of careers. Data from the US General Social Survey indicates that sons of fishermen are 275 times as likely to become fishermen themselves, and daughters of military officers 281 times as likely to become soldiers.[9] Huntington's disease is also inherited, but with a much higher penetrance than fishing for a living. It is caused by a fault in the huntingtin gene (HTT), which encodes a protein that transports things down the axon (the trunk of neurons). The huntingtin protein is also known as IT15 for 'interesting transcript

---

* St Vitus had a very short life – he was killed by Diocletian after driving a demon out of him. Which goes to show, you just can't please some people (or that Vitus failed to get rid of all of the demons).

221

15' (geneticists once again demonstrating a lack of imagination); interestingly, I couldn't find the purpose of interesting transcripts 1 through 14. The HTT gene contains a repeat of three DNA bases (cytosine-adenine-guanine: CAG); this encodes the amino acid glutamine, and the more repeats of glutamine in the HTT gene, the greater the risk of disease. Mutant HTT is, like Aβ, toxic to brain cells, causing damage over time as it accumulates.

Involuntary movements are characteristic of another brain disease, Parkinson's. This chronic degenerative disease received its name after an eighteenth-century British apothecary, James Parkinson – who described it as *paralysis agitans*. Having an eponymous scientific condition seems to be down to chance, and probably not really something to aspire to. Most of the time the condition is something awful, meaning your offspring will be teased for it (no one wants their surname to be a childhood taunt). Alternatively, like poor old St Vitus, you are remembered solely for the disease rather than the many miracles you performed in your twelve years of life. It's probably best to have something named after you that no one remembers except trivia quiz setters; this list includes the saxophone (Adolphe Sax), mesmerise (Franz Mesmer) and remarkably maverick (Tom Cruise – just kidding, it's after Samuel Maverick, who rebelled by *not* branding his herd of cows, slightly less rebellious than flying low over an airfield and playing homoerotic volleyball).

Parkinson's disease presents as uncontrolled tremors, a result of a loss of nerve cells in the substantia nigra. This is found in the midbrain and controls reward and movement through the release of dopamine. It is thus susceptible to many addictive drugs, such as cocaine which blocks dopamine reuptake. The greatest risk factor for Parkinson's is advancing age; other risk factors include pesticide expo-sure, rural living and drinking from wells. Curiously, smok-ing and drinking reduce your risk of getting it – though this needs to be carefully weighed up against all of the other diseases they do cause. Coffee drinking is protective (reduc-ing the risk by about half), but curiously, drinking tea is not. This suggests that the protective compound isn't caffeine but something else. Narrowing this down is tricky, coffee contains over one thousand compounds – marginally fewer than tobacco smoke, but still a mess to pick apart. A compound called eicosanoyl-5-hydroxytryptamide might be the key player[10] and it may also protect against Alzheimer's,[11] so hit the lattes, as long as they don't cripple you financially.

Parkinson's disease can be treated with a drug called L-DOPA (also called Levodopa), a precursor molecule of dopamine. Marcus Guggenheim isolated L-DOPA from the broad bean plant. He didn't patent it because he didn't think it had any useful activity, coming to this conclusion after trying it on himself, leading to explosive emesis. The lesson as always – self-experimentation with homemade drugs can have disastrous consequences. Oliver Sacks used L-DOPA in the 1970s to treat patients with sleepy sickness* (*Encephalitis lethargica*). His patients had been unable to move since the 1920s; the remarkable turnaround the drug caused was captured in his book *Awakenings* and the film of the same name, starring the late, great Robin Williams.

Multiple sclerosis (MS) also affects movement. It is normally diagnosed at a relatively young age, but adults

---

* Not to be confused with *sleeping* sickness, a trypanosome (para-site) infection spread by the tsetse fly.

with an older onset age tend to get worse faster. MS is an autoimmune condition, sharing features with other diseases where the immune cells start attacking our bodies, such as diabetes, which we will look at in more detail in the next chapter. In the case of MS, the immune system attacks the myelin that surrounds the nerve cells. This reduces the efficiency and speed at which signalling can occur, both in the brain and the peripheral nervous system, leading to muscle weakness and co-ordination troubles. Approximately 100,000 individuals live with MS in the UK (enough to fill Wembley Stadium). Like other autoimmune conditions, it is more than twice as common in women than men.

The last brain condition associated with ageing we are going to look at is epilepsy. Over 600,000 people in the UK live with epilepsy – nearly one percent of the whole population – with nearly fifty million people worldwide. One in four people diagnosed with epilepsy are over sixty-five. Epilepsy is characterised by seizures[*] – sudden, uncontrolled bursts of electrical activity in the brain. There is a range of symptoms from confusion to loss of consciousness and jerking, uncontrollable movements of arms and legs. Most people visualise epilepsy in its most severe form, causing people to pass out and start twitching. But these kinds of generalised tonic-clonic (formally *grand mal*) seizures account for only about ten percent of all epileptic events. About half of later-onset epilepsy cases are associated with some other form of brain damage, be that stroke, brain injury, tumour or dementia, but for the other half it is yet

---

[*] The term 'fit' is no longer widely used because of its somewhat derogatory connotations.

another brain condition that falls into the idiopathic* box. Various things can trigger epileptic seizures including tiredness, stress, alcohol, skipping meals and fever. Flashing lights *can* trigger epilepsy (photosensitive epilepsy), but only rarely, accounting for about three percent of people.

ZEN AND THE ART OF BRAIN MAINTENANCE

As with other ageing conditions, medicine cannot reverse cognitive contraction. The best we can hope for is to delay the decline, or at the very least not to accelerate it. There are constant little cues that my mental sharpness has lost some of its lustre – more often than not when trying to recall the name of someone in the film I am watching who appeared with the other person in that show I just watched. I suspect some of this decline is mobile phone-linked. And not in a '5G will give you COVID' way, but in an accessibility of knowledge way. If I can access all of the world's information at my fingertips, using storage space to remember that Alyson Hannigan starred in both *Buffy* and *How I Met Your Mother* feels wasteful. It's also not clear whether remembering that her character was called Lily, and what she did at band camp in *American Pie*, takes up space I need for more important things – such as children's names, friends' birthdays or where I put my flute.

There is an alluring school of thought that compares the brain to a muscle, which through training can be kept in tip-top condition, leading to a slew of 'brain training' apps of variable value. A problem with evaluating the impact of brain

* The medical way of saying we haven't a clue, without saying we haven't a clue.

training is that the thing being measured – memory/cognition – is a bit vague compared to other more defined outputs (e.g. heart attacks) and therefore requires huge numbers of volunteers to draw valid conclusions. One approach to dramatically increase study size is to undertake 'citizen science' projects. Researchers tried this in conjunction with the BBC show *Bang Goes the Theory*; over eleven thousand people were tested with an online training app. Study participants did indeed get better at the tasks they trained to do, but this did not transfer to other tasks.[12] In essence, training to put the right shape in the right hole makes you faster at doing that task but won't help you remember that Count Olaf was also Barney Swanson, or how I made a reference to the same show one paragraph ago. Repeating acts reinforces specific neural pathways, but dementia causes random pathway loss; the ability to play a particular game well won't stop damage elsewhere. As with many things in academia, the value of brain training is a cause for heated discussion.* For example, in one paper I read, the authors concluded, 'There is no convincing evidence that working memory training is NOT effective';[13] which isn't quite the same as stating brain training *is* effective. Whilst it didn't begin with 'that's you that is', this passive-aggressive double negative, penned in response to another article, is an excellent example of the fine tradition of drawn-out arguments in academic literature, conducted through letters to the editor. Sadly, the long-form scholarly

---

* To emphasise this point, Henry Kissinger once said, 'Academic politics are so vicious, precisely because the stakes are so low.' The quote is itself a battleground for such pettiness, having also been attributed to Woodrow Wilson, William Sayre and Sayre's colleague Richard Neustadt. Proving academics will always find things over which to fight.

squabble, serialised over several issues of niche journals, is a dying art thanks to the immediacy of social media; bah humbug.

The consensus is that whilst brain training is probably a good thing, or at least not a bad thing, most of the apps that offer 'brain training' lack scientific validation. Which means that my guessing five-letter words and getting irate when trying (and failing) to beat my friends isn't detrimental but is unlikely to change any health outcome. Of course, brain training isn't restricted to online games: board games in general and chess in particular have all been associated with reduced risk of dementia.[14] As have many forms of 'leisure activities',* including reading, playing a musical instrument and doing crossword puzzles; the most protective hobby was dancing – probably because it combines cognitive, social and exercise elements. Even if there isn't a direct biological link between gaming and health, the social connectedness from playing will be beneficial, unless you play Ludo, in which case you deserve everything you get.

If the idea of flipping over the Monopoly board seems too traumatic a way to stop memory loss, much is made of the protective value of meditation (or mindfulness). Especially by those who do it (and sell apps/retreats). But the associations are again weak; in one study, the physical structure of participants' brains was assessed before and after an eighteen-month course of meditation. It made no difference, though some benefits accrued relating to behavioural outcomes of attention

---

* Curiously, the study authors included doing housework under the category of leisure activity. Even more weirdly, frequently climbing stairs was associated with increased dementia risk – this has a strong *correlation not causation* feel: having to return upstairs because you forgot your glasses, again.

and self-knowledge.[15] Another meta-analysis that scanned 17,800 other studies and did a detailed analysis of forty-seven clinical trials encompassing 3,300 people found meditation moderately reduced anxiety and depression but found no evidence for impact on mood or substance abuse.* As with research on brain training games, finding tight associations between meditation-type interventions and their beneficial impact is tricky. The studies are by necessity long, the outcomes are tricky to define and there are few measurable markers to help the numbers. They also face a challenge common among behavioural studies: how to create a blind study group given that the treatment is a self-directed and internally defined behaviour – in this case meditation? To give a comparator of the relative simplicity in my field of vaccines, we can study whether a vaccine works by testing whether vaccinated volunteers get infected with a specific pathogen compared to unvaccinated controls; the binary outcome (infected or not) makes studies somewhat more straightforward; as does the ability to use placebo injection controls. Even then the trials are massive (the Moderna RNA vaccine trial had thirty thousand participants) and not without their critics!

FEEDING THE INFLAMED BRAIN

One proposed benefit of meditation is stress reduction. Since stress, cortisol and inflammation are all linked, lowering stress through meditation could lessen its ill effects. More broadly, the thread of inflammation woven through

---

* Mindfulness and meditation have subtle differences, but they are close enough that grouping them into one big *om*-humming meta-analysis can increase the robustness of the data generated.

all diseases of ageing suggests that 'reducing inflammation' could be protective against what does us ill. Which also falls into the 'easy to say, hard to do' category. There are clearly things that increase inflammation – smoking, drinking, steak – but beyond cutting these out, reversing the trend comes with lots of questions. How to reduce inflammation? Should you reduce it in a single organ (the brain) or the whole body? Could reducing it too much put you at risk of infection or cancer? One approach is medication. We have access to a range of anti-inflammatory drugs, from every-day ones such as ibuprofen to steroids, metformin (the anti-diabetic, potential 'miracle drug', more of which in the next chapter), and targeted antibodies.

But taking more pills doesn't seem the best way, and many people argue that you can reduce inflammation through dietary change (again, often those with diets to sell or social media accounts to promote). The question is whether diet can reduce inflammation in a meaningful way, with the corollary of whether specific foods have any proven impact on the development or progression of specific diseases. The shallowest of skims on the internet will provide a shopping basket full of anti-inflammatory 'super-foods'. The recurring items on this list would make an excellent middle-class brunch: colourful foods (berries, tomatoes), fatty fish (salmon), green tea and avocadoes. Apart from cost, their other unifying feature is colour – turmeric is widely touted as a miracle food; saffron to a lesser extent, presumably because good saffron costs so much. It is far, far easier to find lists of superfoods than solid epidemiological data supporting their health value; and it is nearly impossible to find mechanistic data showing how eating colourful foods alters your immune system.

Except as we saw in the well-argued case for tomatoes I made in Chapter 1.

As appealing as it may sound, the idea that taking a nutrient capsule or consuming immense quantities of avocado/açai/kale to cure all that ails you is almost certainly wrong. The systems of the body are remarkably complex; disease is a combination of genes, environment and luck and therefore eating all the blueberries would still not fix smoking twenty cigarettes a day. Much of the research identifying the benefits of individual food components can be viewed as the inverse of research into the carcinogenic nature of substances. Both use mouse models, dosing Mickey up with enormous doses of a particular compound de jour and then extrapolating like crazy. Don't get me wrong, animal studies are an important step on the way to discovery, but drawing a straight line from them to human immortality doesn't work. This extrapolation problem particularly applies to supplements, where the amounts needed are several orders of magnitude larger than what can be reasonably obtained by eating food. Taking turmeric as an example, to cause an effect, an average person needs to consume one thousand milligrams of the active ingredient (curcumin) daily, which doesn't sound that much (it's only one gram after all). But to get that amount you need to consume two and a half teaspoons of turmeric every day. That's a lot of yellow food.

Much of the beneficial colour in our food comes from plant compounds. These come in a range of hues: anthocyanin makes blueberries blue, lycopene makes tomatoes red, carotenoids make carrots orange (or purple). At one stage in my PhD I went down a brief rabbit hole of looking at whether we could artificially make lycopene (the red in tomatoes); I am still not entirely sure why. These biochemical compounds, particularly flavonoids, can act as antioxidants, soaking up chemicals called reactive oxygen species which can damage DNA. Antioxidants are widely pitched as having health benefits – with one Dutch company even going so far as to suggest they provide a medicinal boost to cannabis! But large clinical trials of antioxidants on their own showed no benefits. As Professor Gary Frost, Chair in Nutrition and Diabetics at Imperial, told me: 'We know that eating more fruits and vegetables is beneficial, but if we break them down into individual components, we don't get the same benefit.'

FOOD FOR THOUGHT

Rather than looking to individual superfoods, it is better to consider the diet as a whole. The most widely accepted dietary pattern for health benefit is the Mediterranean diet, first identified by Ancel Keys, American physiologist, adventurer and polymath.[16] Born 1904 in Berkeley, California, Keys worked during his childhood in a lumber camp, a gold mine and a cave, where he was employed in the glamorous task of shovelling bat guano. During his undergraduate years he worked as an oiler on a steamship. He then got two PhDs (I don't quite understand why, one was bad enough for me) before studying the effect of high altitude on his own blood by living above 20,000 feet in the Andes for ten days. His first major contribution was the development of the K-ration.* Keys developed this staple of the US Army's

---

* Which sadly, though it begins with the same letter as Keys, probably received its name to distinguish it from other ration types (A and B): the United States Quartermaster department isn't renowned for its imagination. See also Item #9118100, Prophylactic, Mechanical, Individual, 144 or the GI Jonny.

Second World War efforts by purchasing three thousand calories of assorted packaged goods that would fit in a paratrooper's smock pocket in a local supermarket (history doesn't record whether he visited a K-Mart). Whilst effective, it's fair to say these weren't wildly popular – being described as 'palatable' and 'better than nothing' by the first six test cases.

Keys then moved on to perform a remarkable – probably unrepeatable – study on starvation. The aim of which was to understand how to start refeeding the millions of displaced people and POWs at the end of the war. Thirty-six conscientious objectors volunteered for the study; their energy intake was reduced to 1,800 calories for three months. The volunteers lost nearly twenty-five percent of their body weight and understandably became obsessed with food. A rehabilitation stage followed; the main finding was that as long as the recovery food contained a lot of calories it didn't really matter what form it took.

In the context of our story on ageing, Ancel's key impact was investigating the interplay of diets and heart disease. He set up the Seven Countries Study, comparing twelve thousand healthy middle-aged men from Italy, Greece, Yugoslavia, the Netherlands, Finland, Japan and the United States. The underpinning idea was that post-war dietary changes in the US increased the frequency of heart attacks. The study observed a tenfold higher rate of heart disease in Finland than in Crete or Japan, and Keys surmised that the countries with lower fat diets had lower rates of cardiovascular disease. This led Keys to propose the benefits of a 'Mediterranean diet', one rich in fruits, vegetables and olive oil with lower levels of saturated fats. The benefits of the Mediterranean diet have been debated, digested and refined since Ancel's original study,[17] and two key points emerge: replacing saturated fats with polyunsaturated fat is beneficial, as is focus on overall dietary patterns rather than single nutrients.

Which means that simplistic advice like 'five a day' for fruit and veg, or 'eat the rainbow' can be effective. And the messaging works: in a 2023 OECD survey, the UK was one of the top countries for consumption of five fruit and veg a day.[18]

SLEEPY TIRED PEOPLE HAVING NAPS

Having simplified the complex arguments about diet, inflammation and dementia down to 'eat more veg', there is one other lifestyle change that needs considering: sleep. Within the complex world of understanding the brain, sleep is one of the most complex. I spoke to Professor Bill Wisden FRS, sleep expert, to get some insight. The answer to the question 'what is sleep?' was simple: 'It is a reversible state of unconsciousness.' The answer to the question 'why do we sleep?' was also simple, but slightly surprising: 'We don't know.' And Wisden is one of the world's leading experts on sleep. I was impressed by his candour.

His response was surprising because sleep is such a core part of life – we spend about a third of our lives asleep, twenty-six years on average! Wisden went on to explain that all animals sleep, even fish, and that it is a fundamental drive like hunger. Everybody hurts without it. The longer you lack sleep, the more sleep you need to catch up again. But it's not the end of the world, and you *can* catch up. A night's kip has two distinct stages: rapid eye movement sleep (REM) and non-REM sleep. When we fall asleep, we enter the non-REM phase first. The electrical pulses in the brain slow, taking on the delta wave form associated with deep sleep. During deep sleep the heart slows, the body temperature drops and the brain uses less energy. If uninterrupted, the brain then cycles into REM sleep – as the name suggests in this period, the sleeper's eyes move rapidly.* The body also enters a state of near paralysis during REM sleep, stopping you from falling out of bed/off a branch. Dreams are automatic for most people in REM sleep, but you can also dream in non-REM sleep. Whether dreams serve a physiological purpose or are just a by-product of something else is also not known. On some levels, the brain looks conscious during dreaming, but as it makes fewer of the key neurotransmitters needed to lay down memory, consequently the dreams fade. The whole process takes about ninety minutes, meaning in an eight-hour sleep you might cycle through the different types four or five times.

The dawn of our understanding of sleep comes from Baron Constantin von Economo, born in Brăila, Romania, in 1876. Constantin was another turn-of-the-century polymath – one of the first people to learn to fly, he served as a pilot for the Austro-Hungarian Empire during the Great War. His major contribution came from exploring sleepy sickness (*Encephalitis lethargica*) – the same disease that Oliver Sacks described in *Awakenings*. Influenza virus infection almost certainly caused the disease (it occurred around the time of the 1918 pandemic), damaging the brain – a kind of long flu in modern parlance. His patients developed either insomnia or narcolepsy, and he identified that

* Underneath the closed eyelids, unless you sleep with your eyes open like a total weirdo!

these contrasting conditions correlated with damage to different areas of the brain. When unfortunate people lose parts of their brain, it can tell us about the function of that region – take for example the case of Phineas Gage, 'a living part of the medical folklore', famous through all popular scientific books for firing a metal rod through his eye and into his prefrontal cortex, altering his personality. Whilst sleepy sickness disappeared as a condition after 1927, narcolepsy, characterised by involuntary sleep episodes, still occurs. As with sleepy sickness, modern day narcolepsy might still be a consequence of influenza virus infection – but, in this case, rather than killing specific areas, the virus induces an autoimmune response to the hormone orexin. Orexin is named after the Greek word for 'appetite', because as well as inducing wakefulness it stimulates food intake. As we will see in Chapter 9 (diabetes), when the immune system attacks regulatory hormones their function goes screwy.

SNOOZING MY RELIGION

Narcolepsy aside, in general sleep is a good thing! Acute sleep deprivation blunts our abilities. In one study, participants were evaluated for their driving proficiency in a simulator after a night's lost sleep or a glass of wine. Both led to a significant reduction in skill – which suggests driving home the morning after a big party is doubly bad. If you skip sleep for a long time, compensatory mechanisms kick in, boosting things like cortisol. Not sleeping for forty-eight hours initiates microsleeps – an involuntary shutdown of bits of the brain, which can happen without the person even noticing. Randy Gardner holds the record for the longest time without sleep: as a teenager in 1964 he stayed awake for eleven days and twenty-four minutes. His record has never been bested, at least in the official arbiter of records (Guinness), because of the risk to health – amongst other things they also don't accept records for alcohol consumption, blinking, speeding on public roads and freckles (it's unclear what danger some of these pose). As a workaround, Maureen Weston from Peterborough set a record for 477 hours in a rocking chair marathon and Wakeful-Weston's record was in turn smashed by Robert McDonald of California, rocking a chair for a staggering eighteen days, twenty-one hours and forty minutes.

These attempts were made without drugs, but a vast range of stimulants can delay Hypnos's embrace. Armies used various methods to keep their soldiers awake during the Second World War; the allies preferred Benzedrine, the Germans meth (which they called *Hermann-Göring-Pillen*). More recently, speed has been replaced with the slightly safer modafinil, because cranking soldiers up on whizz is now considered a bit questionable; though the long-term effects of not sleeping, on top of the other stresses soldiers endure, are likely to be unpleasant. Whether sustained lack of sleep can actually kill humans remains unclear, though Allan Rechtschaffen in the 1980s showed that after two weeks of being kept awake rats die of sleep deprivation. Unsurprisingly it is not ethically possible to test the same thing in people.

Sleeping patterns have a genetic component: deleting genes in mice can increase or decrease the length of time mice slumber, but no single gene deletion can remove the need to sleep completely. Sleep is also underpinned by the circadian rhythm, a hardwired, highly conserved twenty-four(ish)-hour cycle of activity in our cells. Sleep cycles link to cortisol, the hormone we first encountered when thinking about stress and the heart; cortisol increases in the morning, getting you ready for the day's adventures. Whilst the body's circadian rhythm can run independently to some degree, daylight plays a critical role in regulating it. Fighting the cycle of day and night is bad for you. An increasing body of evidence indicates that working night shifts causes negative health effects, with a forty percent increased risk in cardiovascular disease[19] and an association with breast and prostate cancer.[20] The tight links between cortisol and the circadian cycle means chronic sleep deprivation closely overlaps with stress.

Sleep patterns change with age.[21] The total sleep time decreases from a peak of ten and a half hours in adolescence by eight minutes a decade; we also slowly lose the ability to return to sleep after waking, which happens more frequently too. This reduction represents a biological baseline, but additional factors shorten slumber: stress, distress, medical conditions, needing to pee. In my experience, most changes happened when I was a new parent and my sleep patterns never quite restabilised. Even before children I was a light sleeper and, as my wife will contest, quite often experience 'sleep madness', waking up thinking I have died in my sleep or not knowing who I am; this gets even worse when I sleep in a new place with occasional bursts of somnambulance, once unclad. Hopefully my crazy dreams just reflect an active imagination, but one study suggests an increased risk of early-onset cognitive decline in people who experience intense dreams during childhood.[22]

My weird dreams aside, ageing shifts circadian rhythms, moving them a bit earlier relative to the light cycle. Circadian rhythm is also linked to our genetics (the so-called clock genes), one of the more unlikely things in my 23andMe profile being the prediction that I am likely to wake up at exactly 7:42 a.m. This was based on 450 genetic markers and takes into account age and sex, and they mention that the predictive power of the statistical model (the R^2) was pretty weak – 0.16 (on a scale that goes 0–1, where 1 is good). That said, I woke this Sunday morning at 8:00 a.m. and when I need to get up an hour earlier for work I tend to struggle. The time of day affects everyone, but our genes determine the way in which our bodies and critically our minds navigate the 24-hour cycle. Chronotype, or diurnal preference, is driven by mutation in the clock genes;[23] there really are Larks and Owls. Whilst a day is twenty-four hours, without the guidance of the sun and other cues (*zeitgeber* or time-giver in German), our body clocks reset to a genetically determined value which can vary from 24.3 hours in owls to 24.1 in larks.

Either way, I can predict I will take a post-prandial nap today, as is my right as a father of a certain vintage. One of the least surprising pieces of research that I read whilst writing this book is that napping frequency increases with age.[24] The good news (for me at least) is that having a quick nap in the afternoon can be beneficial. Naps between five to fifteen minutes can immediately give up to three hours of improvement on cognitive tasks – for example writing a book section about sleeping. And there is some evidence napping might even protect against dementia – at least that is my excuse next time my colleagues catch me asleep in my office during an after-lunch meeting.*

---

* Another colleague noted that I had now transitioned to full professor because I appeared to sleep through her seminar and still ask a relevant and challenging question at the end; a core academic skill.

We need to ask whether lack of sleep causes dementia or sleep disruption is caused by dementia. Damaged synapses don't only appear in memory areas of the brain; dementia affects all functions and therefore can affect sleep. There is a correlation: the Whitehall II study looked at a cohort of nearly eight thousand British civil servants recruited between 1985 and 1988.[25] Those who reported less than six hours sleep a night in their fifties and sixties had a thirty percent increased risk of dementia. It didn't take much to reduce risk; those who reported seven hours a night had a lower risk of dementia.

I set about to improve my sleep as a way to protect my mind. Getting seven hours sleep a night feels like it should be a relatively easy hack. The problem of course is the circularity of insomnia and adding the worry of long-term effects isn't necessarily going to help rock you to sleep. The night I started recording my sleep, I got stuck in a one-hour loop of semi-sleep trying to invent a game that combined elements of Betrayal at House on the Hill with the garage band The Streets (this may or may not have been because I had eaten blue cheese, drunk red wine and played the game whilst listening to Mike Skinner's hypnotic spitting before bed). I looked into using the slumber tracker function on a wearable device but found that wearing a watch whilst sleeping stopped me sleeping. However, I could access three years of data from my wife. Whilst her resting heart rate data indicated when she was unwell (by going up), sleep data had no equivalent patterns. The same applied for me when I kept a sleep diary. There are simply too many variables: room temperature, food, drink, caffeine, noisy neighbours, stress, illness, restless children. One big sea change in recent years has been my teenage children turning in after me; beneficial for my sleep patterns, if not theirs.

Like most people I experience bouts of insomnia, linked to work, family and other stresses. I find the worst thing is being woken thirty minutes into sleep. Reading helps clear the mind. Other simple tips include: go to bed at about the same time each night, sleep in a dark room and take exercise in the day – but not too close to bed. Caffeine produces mixed effects, some people (like me) can't even look at a cup of tea after lunchtime if they want a good sleep, others (my parents) can drink an espresso just before bedtime. I'd assumed the action of caffeine would be well established, but it was another one of the questions to which Wisden said we just don't know the answer. The complexity of the brain makes it hard to know what needs to be improved or measure what is changing. 'If only it was that easy,' sighs Wisden.

One of the modern problems is blue light from our handy mobile devices. Blue is particularly disruptive – because it is the colour of the sky and therefore tricks our brains into wakefulness, buggering up the circadian rhythm. Doom scrolling social media on your phone therefore provides a double negative hit – the light waking our bodies, the endless cycle of bad news stressing our brains.

Within the murky world of people selling anti-ageing products, people selling things to help improve your sleep occupies a particularly shady spot. Some make sense – better mattresses for example. Some don't – for example, sheets grounded via an earth wire to reduce electrical build-up while you snooze, which apart from providing me with the chance to make a pun about stat-kip electricity cannot possibly be of any use. The strangest suggestion I have seen is to sleep with your mouth taped up, forcing you to breathe through your nose. I would last exactly one breath before a massive panic attack ensued. The best suggestion (and admittedly no more practical than any of the others) is to acquire a second bedroom to which you can retire if either you or your partner can't sleep.

Whilst the associations between stress, sleeplessness and dementia are relatively weak, if you have been paying attention the main risk factors should hold no great surprises[*] – especially if you consider much of dementia is vascular (related to blood and circulation). The things that damage your heart also damage your brain. There is of course the chance that cardiovascular risk factors will kill you of something else before the dementia kicks in, but more likely the heady cocktail of bad diet, booze and fags will lead to an equally heady cocktail of heart failure, COPD and dementia.

HEARING IS REMEMBERING

Interestingly there is one, final, unexpected dementia risk factor: hearing loss. Both preventing it in the first place and, if necessary, correcting it. The second part is reassuring for middle-aged me, as young me did their very best to ruin my hearing. A combination of concerts and in-ear headphones certainly didn't help. But the real killer was machine-gun fire, especially when shot in an enclosed space without ear defenders on. You may not know that you needed to know this, but a 5.56 mm calibre round produces about 160 decibels (dB) when fired, which puts it 30 dB above the threshold of pain, and about 90 above the safe limit. For those of you

---

[*] High blood pressure, lack of exercise, smoking and drinking all increase risk of dementia.

who haven't had the (dis)pleasure of listening to rifles in close proximity, they are louder than jet engines; all told I was pretty stupid to forget my ear plugs. It inevitably led to a long period of tinnitus, which, if you have fortunately avoided it, presents as ringing in the ears. Even if you don't expose your ears to unduly loud noises, hearing declines with age. The hairs that detect sounds begin to fail, particularly high-frequency sounds. Uncorrected hearing has a linear relationship with dementia – the more severe the hearing loss, the more likely dementia.[26] However, the strength of the association needs further exploration. Several mechanisms have been proposed, including cognitive decline due to lack of input and social isolation. The good news is that it is correctable: when hearing is restored with hearing aids, the risk of dementia dramatically declines.

All this thought about thinking has made me hungry, and what could be better than a tasty snack? But sadly, uncontrolled glucose has been linked to vascular dementia. The main cause of this is diabetes, killer number 6. We thus delve deeper into our abdomen to explore the role of the pancreas and its vital product – insulin.

CHAPTER 9

Don't Sugar-Coat It: Killer Number 6 – Diabetes (and Autoimmunity)

'Let food be thy medicine, let medicine be thy food.'

Hippocrates

Like it or not, sugar fuels your body. You may consume this fuel in a different, more rarefied form – low GI, brown pasta with a side of steamed vegetables (all beautifully presented on your social media) – but your body then smashes it down into glucose. Glucose acts as the universal currency for all cells. In order to extract energy from glucose, our cells contain tiny little bacterial passengers called mitochondria, the cell's powerhouses. They 'burn' sugar in oxygen in a process called respiration, releasing energy in the form of another chemical, ATP, which the rest of the cell then uses to go about its daily life. The resultant energy created from burning glucose in oxygen is critical for life. Without energy, your body dies. Without oxygen to burn the glucose efficiently, your body dies. Some tissues can briefly metabolise glucose without oxygen but, a bit like burning wood without enough air, you get less energy and more filthy by-products – in our case lactic acid.

The life-giving glucose is carried around the body in the blood, specifically in the liquid part, plasma, which we first encountered in Chapter 5 (strokes). But in the case of blood glucose, you can definitely have too much of a good thing. And so enters candy-coated killer number 6 – diabetes. Too much sugar in the blood is called hyperglycaemia (from the Greek *hyper*, for 'over'); too little is hypoglycaemia (*hypo*, as you can guess, being Greek for 'under'). When measured as an average across the day, the concentration of blood glucose is tightly regulated, with peaks and troughs following food intake, explaining why it is often assessed after an overnight fast. A healthy fasting glucose level in the blood is 100 mg/dl ('dl' being one decilitre – one-tenth of a litre of blood). Which in an average-size adult works out as about five grams of glucose in total – basically a teaspoon's worth. We can dramatically change this by eating. As an example, the 10.4 g of sugar in the Tunnock's caramel wafer I just ate (one of one million bars made every day) will take approximately fifteen minutes to start entering my bloodstream.* Assuming I have five litres of blood and it completely absorbs the bar's sugar, my blood glucose will increase by 200 mg/dl – to three times healthy levels. But it will only stay at this elevated level for a short time, as the body rapidly reacts to fluctuations in blood glucose, which is where insulin, produced by the pancreas, comes in.

The first indication of the involvement of the pancreas in glucose regulation came when Oskar Minkowski, a German physiologist, removed one from a dog in 1889 and discovered glucose in the canine urine, the pity being that he didn't work on bears so I could use the phrase ursine urine. Subsequently, an American doctor, Eugene Opie, narrowed

---

* Not to mention the teaspoon of sugar I stirred into my tea to wash it down.

pancreatic glucose control to a specific region, the islets of Langerhans, named after a third scientist, Joseph Langerhans, who discovered them during his PhD. Various scientists then ground up bits of dog in an attempt to isolate the key compound, with two different people independently coining the word insulin (from the Latin word *insula*, meaning 'island') after Langerhans' pancreatic archipelagos. A Canadian, Frederick Banting, first isolated insulin. Banting's moment of inspiration occurred at 2 a.m. on 31 October 1910 when he conceived a better way to isolate the protein. The First World War interrupted his research, though he did receive a Military Cross at the Battle of Cambrai. After the war, Banting worked in John Macleod's lab at the University of Toronto alongside a lab technician called Charles Best. The team started with dogs, though there were apparently very few strays in Toronto in the early 1920s. They progressed from dogs to calves, tying up their pancreatic ducts and extracting what remained in the islets. For this work, Banting shared the Nobel Prize with Macleod in 1923, making him the youngest ever winner. The technician Best didn't get the Nobel Prize, but Banting did give part of his prize money to him. Best's participation was fortuitous; he received his appointment on the basis of a coin toss ahead of another technician, Clark Noble, and because of this Noble missed out on sharing in the prize money of a Nobel. This may seem like one of the most unfortunate coin tosses of all time, but remarkably, the UEFA Euro 1968 semi-final was decided by the flip of a lira – Italy nailing it by calling tails, which puts losing to them on penalties in the 2020 final (actually played 2021) into perspective.

Insulin increases glucose uptake from the blood into cells, laying it down for future use. Glucose can be stored either as a large sugar macromolecule called glycogen, made up of multiple glucose sub-units strung together, or as fat. Insulin drives this energy deposition by binding to a protein called the insulin receptor, found on the surface of cells in the liver, muscles and fat tissue. Muscle cells tend to store glucose in the glycogen form; fat cells make fat and the liver does a bit of both. Insulin is produced within the islets of Langerhans, specifically by the beta (β) cells which store it in little granules awaiting release (Figure 17). β cells act as glucose sensors, constantly reading levels in the blood; when glucose rises, the β cells spit out their packages of preformed insulin to act across the body, restoring the level. This all happens extremely quickly – within ten minutes of glucose spiking – meaning that half an hour after eating a delicious milk chocolate-coated wafer biscuit my body is already bringing the sugar rush under control.

But insulin only makes up half the story. Whilst I can spike glucose levels with ease by eating red- and gold-wrapped sugary treats, making them dip is harder. Having saved the energy for a rainy day, we need a way to get it back. For example, my bitterly cold ten-minute sea swim this morning would have burnt about the same number of calories as the five layers of wafer and four layers of caramel that I consumed, but would not greatly dent my blood glucose levels below their baseline. In other words, the big spike up after eating is not mirrored by an equal and opposite crash upon exercising. Some of this can be explained by the storage of glucose in the muscles as glycogen, giving them a depot for future action and ensuring the muscles don't directly pull the glucose out of the blood (at least initially). The islets also produce an opposing hormone to insulin called glucagon, but in the alpha (α) cells rather than

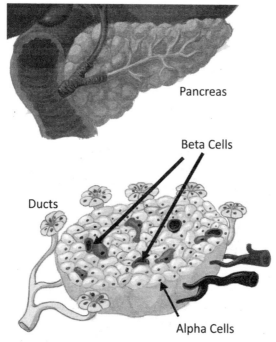

Figure 17. The pancreas. Amongst other things, home of the all-important alpha (α) and beta (β) cells that control glucose levels in the blood.

the β ones. It acts as the yin to insulin's yang, signalling to cells to break down glycogen into glucose. Several different factors control glucagon release, including the stress hormone cortisol – yet another reason why prolonged stress can profoundly and negatively impact health.

HONEY I SWEETENED THE PEE

For ninety-five percent of the UK, blood glucose regulation happens without thought or consequence; however, nearly 4.8 million people in Britain and 422 million people worldwide live with diabetes mellitus. Named after a combination of the Greek word *diabetes* meaning 'flow' and the Latin word *mellitus* meaning 'honey'. It received this name because, when untreated, diabetes causes the urine to be sweet (hence Osler's observational prank in Chapter 3). The ancient Egyptians first noted the sweetness of diabetic urine in the fifteenth century BCE, but it took an Englishman,* Thomas Willis, to taste it and add the word 'mellitus'.

Endocrinologists split diabetes into two major types (called type 1 and 2; sometimes T1D and T2D). Type 1 diabetes accounts for about ten percent of cases and mostly occurs early in life; it is characterised by the destruction of the β cells of the pancreas by the immune system. This stops the production of insulin. Of the two, T1D is the more severe condition – because it is irreversible and has an earlier onset, requiring people to manage it for longer.

However, the more concerning type for someone my age is type 2 diabetes. Rather than death of the insulin producing β cells, type 2 diabetes is characterised by insulin resistance. The muscle, liver and fat cells no longer respond to insulin in the blood. Lifestyle factors, predominantly obesity and poor diet, are the major drivers. It also strongly associates with ethnicity: people from South Asia (India, Bangladesh and Pakistan) have a sixfold higher risk of T2D. Type 2 diabetes has a relatively slow onset, with a gradual increase in blood glucose from a normal fasting glucose level of 100 mg/dl; anything above 125 mg/dl is considered diabetic.

* Back when we were Top Nation and before History came to a . (see Sellar and Yeatman).

SWEET RELEASE

Both hypoglycaemia (glucose too low) and hyperglycaemia (glucose too high) are dangerous. In the short-term hypoglycaemia is more problematic because of its rapid onset and serious consequences. Blood glucose below 70 mg/dl can lead to blurred vision, drowsiness and, if uncontrolled, seizures and coma.

Hyperglycaemia is a longer and slower killer. Because our blood washes around all of our tissues, diabetes causes damage throughout the body, predominantly to small blood vessels at the extremities. To understand how, I spoke to Nick Oliver, professor of human metabolism and consultant in diabetes and endocrinology at Imperial. We met outside the Cambridge wing of St Mary's hospital where we both work (and Nick studied). As you'd expect from an endocrinologist, Nick is depressingly healthy and cycles everywhere. He told me that effectively every serious problem associated with diabetes has blood vessel damage as its root cause. The excess sugar decreases the elasticity and the bore of blood vessels, reducing the amount of liquid that can pass through them in a way not dissimilar to excess salt or excess fat (and they are all connected, diabetes also causes hyperlipidaemia – too much blood fat). The net result of diabetic vessel damage is that capillaries stop working and the tissues they were feeding die. This leads to a cocktail of unpleasantness including: nerve damage in the hands and feet (peripheral neuropathy), damage to eyes causing problems with sight (diabetic retinopathy) and an inability to fight off infections, particularly in the skin because immune cells fail to reach the site of infection. These processes are called microvascular damage: damage to the eyes is a particular tell of diabetes. Oliver put it eloquently: 'The eyes act as the windows to the glucose soul.'

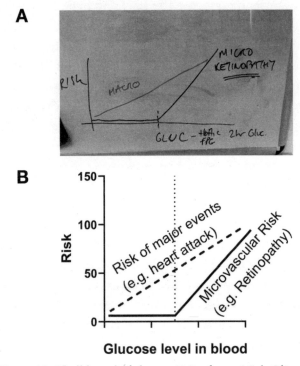

Figure 18. Chalkboard of doom. A) Professor Nick Oliver's original drawing of risk of disease when blood glucose increases. B) Graph of the same.

As the diagram Nick drew on the whiteboard in my office 'very clearly'* shows, the risk of microvascular events (the bendy line) has an inflection point (Figure 18). Up to a certain level of blood glucose, there is essentially no risk, above it risk increases in a linear fashion.

* Nick nicely demonstrating the stereotype about doctor's handwriting.

In addition to causing more bespoke problems, diabetes is a contributing factor to heart disease, dementia and stroke by damaging the blood vessels that feed the heart and the brain. This kind of damage is known as macrovascular and is illustrated by the straight line in Oliver's graph. If you factor in the overlapping risk factors for type 2 diabetes with heart and brain diseases – diet, obesity, lack of exercise – you can see how they contribute to a negative feedback loop of risk. This explains why macrovascular disease doesn't have the same risk inflection point as microvascular damage.

Treatment options for the two conditions differ, reflecting their different aetiologies; type 1 is caused by a lack of insulin, type 2 by a failure to respond to it. One hundred years ago, T1D effectively carried a death sentence, with a one hundred percent mortality rate; affected young people just wasted away. The discovery of insulin by Banting and Best opened up treatment options. By 1922 they had purified enough calf insulin to treat a patient, Leonard Thompson, and it proved a miracle cure. Thompson went from being too weak to hold up his own head to walking independently. Insulin treatment extended his life by thirteen years. Banting and Best sold the resulting patent rights to the University of Toronto for $1, with Banting nobly stating, 'Insulin does not belong to me, it belongs to the world.'[1] Treatment with insulin has been refined since then. A major step forwards was the ability to genetically alter the bacteria *E. coli* to produce biosynthetic human insulin in 1978. Many people living with diabetes now have continuous monitoring with insulin injections from the same device – effectively an external artificial pancreas.

People with type 1 diabetes lack insulin because their immune systems are in overdrive and their T cells kill the insulin-producing β cells in the pancreas. This self-inflicted damage puts T1D in the category of autoimmune diseases.

YOU DO IT TO YOURSELF

Our immune system evolved to prevent us from dying of infection. Sometimes it gets a bit overzealous and instead of attacking things invading the body, it attacks the body itself. Autoimmunity is predominantly (though not exclusively) a feature of the adaptive immune system (the bit that can learn from experience). Autoimmune T and B cells for one reason or another misidentify a human protein and start attacking it. This is bad enough, but the problem is that, unlike an invading pathogen, human proteins don't go away and the immune onslaught intensifies. The signs and symptoms of autoimmunity relate to the part of the body under attack. For example, when the immune system damages β cells, reducing the ability of the pancreas to make insulin, it causes diabetes. But if T cells target the myelin that lines the nerve cells, affecting the ability to conduct signals properly, it causes MS. Autoimmune conditions affect one in ten people in the UK and this is rising.[2] The most common condition between 2000 and 2019 was Hashimoto's thyroiditis (280,000 cases). This was surprising (to me at least), because having worked in immunology for twenty years I've never attended a single lecture about it – arthritis, lupus, diabetes, IBD yes, but not thyroiditis. I don't know how to explain this – there is better PR for some diseases, or possibly they have a more severe health impact. Or are less easy to control.

Genetics plays a huge role in autoimmunity risk, particularly a big cluster of genes called HLA, and we will return to these in the context of transplant. Biological sex constitutes another large risk factor; autoimmunity occurs significantly (1.74 times) more often in women than men, with an even higher incidence for specific conditions such as lupus.[3] Both sex hormones and sex chromosomes drive this association. A trade-off in immunity levels between men and women has also been postulated. Men have slightly less reactive immune systems, making them more prone to infections and cancer; women have more reactive immune systems, protecting them from infection but contributing to autoimmunity.[4]

Sex and genetics set the stage upon which autoimmunity plays out, but they need a trigger. In many cases this will be an infection, though for a lot of conditions the exact agent remains unidentified. Members of the human enterovirus family have been proposed as viral initiators of insulin vanishment.[5] Interestingly, new diabetes diagnoses spiked during the COVID-19 pandemic, and an ongoing project (CoviDIAB) is monitoring cases to determine causality.[6] In addition to its role in causing cancer, another very common virus, EBV[*] has been fingered as the trigger for MS (multiple sclerosis). A huge retrospective survey screened ten million US Army personnel over ten years and found a thirty-two-fold increase in the risk of MS in those with evidence of EBV infection.[7] In parallel, another group identified antibodies in people living with MS that recognised both an EBV protein (EBNA1) and a nerve cell protein (GlialCAM), suggesting that the infection mis-trains the immune response to attack the body.[8] In general, viral

---

[*] The virus that causes glandular fever – or 'mono' – depending on whether you have been snogging or making out.

infections can cause long-term, unwanted sequelae, accelerating the effects of ageing. You face a significantly higher risk of heart attacks and strokes for a month after influenza virus infection.[9] And we can face a whole slew of post-viral syndromes including long-COVID and ME (the specific trigger for which remains unidentified). Viruses are for life, not just for winter.

Causing the body to attack itself is not the sole purview of viruses. In rheumatic heart disease, a bacterial pathogen, group A streptococcus, the bacteria that leads to scarlet fever, plays a role. Most group A strep infections are mild and quickly resolved. In a small subset of people, however, the way that the body fights off the bacteria harms the heart. The received version of how this disease occurs is simple, elegant and, as a result, somewhat incomplete – but it will suffice for our purposes. Our immune system remembers pathogens, enabling it to better fight them off the next time they are encountered. B cells and the proteins they produce, called antibodies, play an important role in this immune memory. Over time and with each bacterial exposure, the antibodies your body makes get incrementally better, improving protection. However, one of the chemicals in strep A bacteria looks a bit like a chemical found in the human heart. In the case of rheumatic heart disease, the highly active antibodies, poised to attack bacteria, bind to the heart muscle instead. This leads to tissue damage, reducing the strength of the heart. (BTW that *is* the simple version, the complex version involves a thing called complement that no one wants to hear about right now thank you very much).

Significant advances have been made in the treatment of autoimmune conditions. A huge breakthrough followed the demonstration that an immune signalling molecule called TNF promoted rheumatoid arthritis. Sir Marc Feldmann discovered this, working with Sir Ravinder Maini who ran the clinical studies. They worked at the Kennedy Institute, at the time based in west London. The institute took its name from Mathilda Kennedy, daughter of Michael Marks, the founder of Marks & Spencer; which is totally irrelevant, but vaguely interesting. Feldmann and Maini's work led to the first monoclonal antibody used as a drug, called infliximab, which neutralises TNF. This class of human-derived drugs is sometimes called biologics. Another biologic targeting TNF, adalimumab (sold as Humira) has made a staggering amount of money, raising an estimated $200 billion in sales.[10] The patent expired in 2023, cheaper alternatives, called biosimilars, becoming available; in 2018 the NHS spent £400 million on adalimumab, but hoped to save £150 million by switching to other biosimilars.[11] As with all things health-related, the figures underlying the NHS drug spend are mind-boggling. In 2021/22 the bill for prescription costs in England totalled £9.69 billion. The most expensive individual drug was elbasvir, the brilliantly successful treatment for hepatitis C virus (£8,250 a dose); the most common item was the cholesterol-busting statin, atorvastatin (53,402,180 doses), and, in spite of costing only £1.50 a dose, because of its almost ubiquitous use made up the largest single expense (£80,560,602).[12]

A better alternative of course is to never develop autoimmune disease in the first place. Vaccines could play a key role in prevention. The link between EBV and MS should provide a stimulus for further research to prevent infections and therefore reduce the long-term after-effects. As more links are made between specific infections and disease types, more vaccines can be developed. Vaccines after all do *work*, really, really well. We should look to a future where we reduce the impact of all pathogens. But pathogens are merely the tip of the iceberg when it comes to our interactions with microorganisms; we live in a microbial world and our interactions with them shape our lives.

KNOWN KNOWNS

The microbes that live in and on us are collectively known as the microbiome. The word can refer to the microorganisms living in any specific habitat, and as such our body's microbiome is composed of numerous distinct microbiomes. The bacterial community (the white gunk) on one of our teeth, for example, completely differs from that found on another; even more amazingly, the bacteria found in one hair follicle differ from the neighbouring one. The ability to sequence DNA cheaply opens the door to this extraordinarily complex world. Estimates vary, but there are at least as many bacteria on and in you as there are cells that make up you. And because of the enormous diversity of these bacteria, the number of bacterial genes far exceeds the numbers of human genes. One of the earliest uses of the term 'microbiota' was in 1956, but it really became entrenched in the scientific consciousness in 2001. By 2023, nearly 150,000 scientific papers on the subject had been published, which seems a lot (the words would fill the Bible one thousand times), but half as many as COVID-19 (380,000 and counting) and far fewer than the heart (1.7 million) or cancer (4.9 million).

The extent and the complexity of the microbiome mean it plays an important role in our health and disease. It has been linked to nearly every medical condition from *Aagenaes* syndrome to zygomycosis.* We definitely need our fellow travellers; they help us break down fibre to get more energy, recover bile and make key metabolites. Mice bred in sterile conditions lacking bacteria are pretty screwy. The microbiome plays a vital role in preventing bacterial infection – which may sound a bit counter-intuitive given it is largely made of bacteria. But they can outcompete other more nasty bacteria, for example *Clostridium difficile*, which causes a common gut infection for people in hospital taking lots of antibiotics. *C. diff* (to give its pet name) exploits this post-antibiotic wasteland. One way to treat it is FMT – a treatment so gross it hides behind an acronym. FMT stands for faecal microbiota transplantation and does what it says on the test tube – bacteria from one person's poop given to another person. The good faeces needs to come from somewhere and you can volunteer to be a donor – and potentially be paid to poop!

Beyond its role in protection against infection, the precise role of the microbiome in disease modulation gets more complex. When asked about one of the microbiome tracking apps, Oliver just sighed. But he did say, 'It is part of us and has a metabolic impact,' before adding: 'most microbiome research is pretty shit, excuse the pun.'

UNKNOWN UNKNOWNS

The challenges of definitively linking the microbiome to health outcomes reflects its complexity. As with all studies

---

* Not strictly true, but alphabetically the first and last diseases I could find.

looking broadly at the human condition, the blurry pheno-types of disease pose a problem – most conditions aren't a single thing, but rather a spectrum that runs from mild to lethal . . . and no disease comes in isolation: dementia corre-lates with cardiovascular disease correlates with diabetes. Then we need to consider the numbers involved; at the begin-ning of the book we looked at the difficulty of linking human genes to health conditions, and we 'only' have 45,000 genes. The microbiome contains orders of magnitude more complexity than the human genome. The whole human microbiome potentially contains 232 million genes. And that's before you include the viruses that feed on the bacteria, which one estimate puts at 10^{13} per person.[13] Our microbes are essentially unique to each of us, or at least sufficiently individual that microbiome-matching has been proposed as a way to trace people, analogous to fingerprints.[14] Even within households there are big differences, though your microbes are more similar to your partner's than your sibling's; which makes sense as (in most communities) we spend more time with our partners and tend to be less intimate with our siblings. And to add a final layer of complexity, the microbi-ome changes constantly. You can change your gut microbi-ome, just by eating something different; the bacteria in my guts after six pints of Cobra and a mutton vindaloo won't be the same as those after a vegan nourish bowl. Food comprises one of a vast spectrum of variables affecting your microbi-ome; a study that tried to comprehensively characterise microbiome diversity found at least sixty-nine different influ-ential factors, ranging from sleep, coffee intake, bread prefer-ence, shift work, constipation and height.[15]

All of which leads to a field of study that can be a bit hand-wavy, and sometimes downright nonsense. It doesn't help that microbiome research can easily be done very badly. The ubiquity of bacteria makes it really easy to find bacterial DNA everywhere, but often it comes from an unrelated, contaminating source. Researchers in a related field of looking at proteins named this the CRAPome (a contaminant repository for affinity purification–mass spectrometry data);[16] and a collaborator of mine coined the 'kitome'[17] to describe bacteria discovered just by doing the study in the first place.* It reflects the computing adage 'crap-in crap-out', only more literally. Equally concerning, as with miracle diets and anti-ageing panacea, often the people promoting microbiome research have some skin in the game financially. Though not always as blatantly as in the paper on the benefits of beer-mediated immunomodulation, published by researchers from the State Key Laboratory of Biological Fermentation Engineering of Beer, Tsingtao Brewery Co. Ltd, who one feels might have a slight vested interest in the positive effects of beer (though they stated no conflict of interest!).[18]

KNOWN UNKNOWNS

Whilst a substantial amount of execrable microbiome research exists, it does not all lack merit. I have been known to dabble in microbiome research, and I hope my data falls in the passable camp. But I am a gifted amateur in this regard and to find out what the microbiome can tell us about health and disease, I spoke to Mr James Kinross, surgeon, author of *Dark Matter: The New Science of the Microbiome* and professional microbiome enthusiast. We

---

* -ome being the catch-all suffice for complex data sets in biology.

did the obligatory Teams meeting, and because he juggles being a surgeon *and* a researcher, Kinross found himself on his third meeting of the day at only 9 a.m. Monday morning. He told me that at its best, microbiome research can help provide a mechanism for how things improve our health. He gave the example of eating fibre – long known to be beneficial, the new link to fermentation by gut microbes to produce beneficial biochemicals underpins how it protects us. However, Kinross does qualify this: 'Microbiome scientists have not really helped themselves, because of the over-hyping of much of the work. We can't possibly meet the expectations that have been set.' Furthermore, Kinross explains, 'There are no standard methods for collection or data analysis,' which means data from one centre can't be compared with another. Even the sample type presents challenges – most sequencing is done on a smear from a poo, but the bacteria on the surface might differ from those found in the middle of the poo. And poo only reflects the final stage of food's journey through our bowels, so faecal flora does not necessarily represent events further up the passage.

One commonly presented view (especially by people who sell them) is of good and bad bacteria. But as Kinross told me: 'You shouldn't anthropomorphise bacteria, if you mistreat your microbes, they will bite you in the bum.' Some bacteria are born bad – if you get a dose of *Vibrio cholerae* you will want a toilet nearby. Others cause problems if they get into the wrong place – particularly if they escape from the guts into the blood. This can lead to the directly adverse effect of blood infection. But misplaced bacteria can also have a more subtle negative impact, increasing inflammation. The immune system is primed to react to bacterial biochemicals and, if found in the blood, they can increase the basal level of inflammation. As seen throughout this narrative, inflammation links to bad outcomes, playing a role in heart disease, stroke and dementia (inflammatory cells shrinking blood vessels); lung disease (inflammatory cells damaging the lung filigree); diabetes (more cells to attack the islets); and cancer (accelerating tumour growth). This process of bacterial bystander inflammation can be further exacerbated by ageing. As we age, our intestinal integrity slowly fails, allowing more bacteria, or at least bits of bacteria, into our bloodstream. This can then lead to further cycle of damage to the gut, inflammation and more bacterial escape.

But outside directly invasive bacteria, many microbes probably just come along for the ride, because the guts are warm and wet and provide an easy source of nutrients – a kind of all-you-can-eat poo-ffet. The exact commensal cocktail for 'healthy' guts remains to be defined. And this uncertainty leads to some fairly woolly phrases such as 'dysbiosis'. If you dig deep enough in shiny press releases, you can get back to the original papers and these suggest a correlation between certain strains of bacteria and health outcomes. The directionality is impossible to discover because the same bacteria are also associated with unhealthy diets.[19] This raises the question – do bad bacteria cause bad health outcomes or do bad diets cause bad health outcomes with the bad bacteria just acting as an indicator? Because more variables can be controlled, researchers use mice to address these types of questions. In an oft-quoted study reminiscent of Joseph's dreams, transferring the gut contents of fat mice into thin mice caused the thin mice to gain weight; remarkably, they didn't actually eat more, they just extracted more calories from the food given. A follow-up study transferred microbes from discordant twins (one fat, one thin) into mice, and again the mice given the 'fat' microbiome put on weight. Extrapolating this into human effects presents challenges – for example, unlike mice, we tend not to eat each other's faeces.

One of the most agreed upon observations is that the greater the diversity the better. Kinross tells me that rather than thinking about specific bacteria, we should consider the function of the whole system. And that having more, different bacteria provides more stability – like in any ecosystem it protects against change. His biggest concern is that thanks to overindulging in antibiotics, humanity might be slowly driving rare species of the microbiome into extinction. Whether WWF would run a campaign to sponsor a verrucomicrobium remains to be seen.

The most commonly cited trinity for microbial meta-morphosis is fibre, pre/probiotics and fermented food. In addition to preventing bowel cancer by accelerating food transit, high-fibre foods offer other benefits. The bacteria in our guts break down indigestible sugars in high-fibre foods, producing short-chain fatty acids (with enticing names like butyrate, valerate and propionate). These compounds are broadly anti-inflammatory and provide a wide range of potentially beneficial outcomes, such as improved gut integrity and reduced airway disease. However, high-fibre diets come with a trade-off, leading to bloating and a somewhat inevitable gaseous output. There is another spurious reason to not eat too much fibre, beyond the flatulence. If you believe one of the inventors of the Corn Flake, fibre can affect your libido. John Harvey Kellogg was a physician who ran a health resort in Battle

Creek, Michigan, and was an early proponent of the role the gut microbiome plays in health. He (alongside brother William and wife Ella) came up with the Corn Flake to counteract the meat-heavy American diet and reduce masturbation. He was not a fan of the 'solitary vice', referring to it in his book *Plain Facts about Sexual life* (rebranded *Plain Facts For Old And Young* in later editions) as a 'sin against nature, with no parallel except in sodomy'.[20] He designed his fibrous Corn Flakes to cure all this bothersome behaviour.

Probiotics are foods (or pills) that contain bacteria; prebiotics are foods that boost the growth of 'good' bacteria. These present a twofold challenge. As alluded to, we don't really know the correct combination of bacteria; just taking some random bifidobacteria in tablet form might not actually change anything. Secondly, unless you have just taken a big course of antibiotics, your guts already teem with bacteria, making it very hard for the newcomers to get a foothold. It's like having a wee in Britain's sewage-laden sea and then suggesting it changed the overall level of pollution. That said, probiotics and live yoghurt should be taken during and after oral antibiotic treatment because they can help re-seed the guts and prevent nasty, antibiotic resistant bacteria (like *C. diff*) taking hold. This re-seeding can have a significant impact. Prior to getting a dose of *giardia** in Turkey, I was somewhat prone to upset stomachs; a dose of Iranian antibiotics followed by Doogh and Mast (yoghurt-based drinks and dips) have given me far greater gut resilience.

---

* A gut parasite – read my other book *Infectious* to see what happened. Not pleasant.

Fermented food makes up the third member of the microbiome troika. These are not dissimilar to pro-prebiotics in that they provide a source of microbes and their beneficial chemicals, but they tend to be more brassica-rich. Fermented foods can have a larger impact on the immune response than high-fibre diets alone.[21] There is no reason not to include them in your diet, unless you dislike garlicky, spicy, rotting cabbage.

Gut bacteria are not fastidious eaters. Yes, they can be carefully nourished on a diet of complex (mostly brown) carbohydrates, but they *love* a bit of sugar, even artificial sweeteners. Russian scientist Constantin Fahlberg discovered the first artificial sweetener, saccharin, mishearing the instructions to 'test' the new chemicals as 'taste' the new chemicals. Or at least that's his story, as chemists love testing their own creations. Albert Hofmann famously took 250 micrograms* of his newly synthesised LSD before trying to bicycle home; on the kaleidoscopic-coloured, hallucination-flecked journey home he decided his neighbour was a malevolent witch; history does not record her response. The inventor of diazepam (Valium) and darling of 1960s mothers, Leo Sternbach, also a keen self-experimenter, stated, 'I tried everything. Many Drugs.'[22] It doesn't always work out so well: another chemist, Barry Kidston, gave himself Parkinson's-like symptoms when trying his home-brew heroin. Even when not deliberately getting high on their own supply, chemists have discovered many useful compounds by accidentally ingesting them: another artificial sweetener, cyclamate, was discovered in 1937 when Michael Sveda took a cigarette break and noticed his

---

* A substantial dose, five times the amount the Drug Enforcement Agency finds in today's acid tabs.

ciggy tasted sweet. The reason it tasted sweet was because Sveda had put his cigarette on the lab bench where it soaked up the chemical. Now, call it health and safety gone mad, but this wouldn't happen today, because smoking cigarettes in a lab full of highly flammable compounds is frankly nuts.

Sweeteners are not completely neutral chemicals – but frankly nothing is. In 2023, the WHO Expert Committee on Food Additives released a risk assessment stating that aspartame was potentially carcinogenic (in group 2B; lower than the risk of red meat).[23] As with many things, you need to ingest insanely high amounts for ill effects. The FDA estimates it would take seventy-five packets of NutraSweet® in a day to exceed the acceptable intake – in sweetness terms the equivalent of six hundred grams of sugar, which would definitely cause more damage.[24] Studies using bacteria and mice suggest that sweeteners, especially saccharin, can change the gut microbiome. As ever, further clinical studies are required, the small amount of research performed so far observed no impact on the human gut microbiome.[25] Ultimately, if they help you eat less sugar and you avoid taking them to crazy excess, sweeteners are probably OK. One exception proves the rule – sugar-free gummy bears, which acquired cult status in the 2010s through a series of narrative online reviews about the 'unholy necromancy' they released on people's bowels.[26]

FIBRE, FERMENTED FOODS AND FLORAL FLURRIES

The majority of our bacterial inhabitants live in our guts. I wanted to see what it would take to change them. One way is through going to space.[27] Sadly, the publisher's advance didn't quite cover the $450,000 required for a commercial space flight (and even then, a longer spell in space is probably needed to see any major change). Instead, I opted for a simpler approach – changing diet. Even that costs money – commercial microbiome kits come in at £300 a pop – yet another example of people being costed out of health. This is where having friends in the trade helps. Dr Kinross and his collaborator Professor Julian Marchesi kindly offered/caved in after my relentless nagging to sequence my stools. My original plan was to eat normally one day, then consume a curry and beer to see the changes. The 'experts' (Kinross and Marchesi) said this would have no effect; but as once observed, we've all had enough of experts, and let's face it they have not witnessed my toilet the day after five pints of Guinness and a Jalfrezi. But in the light of this being science, I let them persuade me to also consume a week-long high-fibre diet, supplemented with fermented food – these being the two dietary interventions with the most evidence for impact.

Step 1 – The 'Normal' Poo. I say normal, but the process ended up being somewhat traumatic. Because of train strikes, I couldn't get to work that day, so I had to collect my sample at home (I still had to self-sample every day – there are limits to the collaborative lengths even Imperial colleagues will go). To do this, I unpacked my Daklapak Fecotainer® – basically a potty that sits on top of the toilet seat. Then commenced a game of battleshits. Attempt one – fail. I missed. Poo in toilet, no poo in the sampler. Attempt two – near miss. Attempt three – nailed it. Prompting the next question – what to do with said sample until I get it to the lab. Answer – the home freezer (double-bagged of course – I'm not a monster*). However, I still needed to transport it to work. This involved

---

* My children very much disagree on this point.

a freezer block, a Lidl cool bag and remembering not to deposit it next to the train's heater on the way in.

Step 2 – The 'Curry' Poo. I was now a pro at the sampling, and I could go at work, making it slightly less traumatic than day one. Though I felt slightly self-conscious carrying a bright blue potty from the loo to the lab. I hid both samples at the back of one of our super cold (-80 °C) freezers. I wrote 'John's shit' on the containers, to avoid confusion and prevent someone accidentally using it for some actual science. The intervention was two bottles of Guinness and a home cooked garlic/chilli chicken; all in the name of science. Though I'm not convinced the intervention was aggressive enough.

Step 3 – The 'Fibre' Poo. Kinross (and the WHO) suggested 30 g fibre a day, alongside three helpings of fermented foods. This turned out to be quite challenging. I filled my shopping trolley with brown food (pasta, bran cereals, bread) and stuff beginning with K (Kimchi, Kefir, sauerKraut). And thus began a long week of counting fibre. To throw some more measurement in, I also recorded stool number, stool consistency, happiness and bloating. I initially chose a forty-three-question-long intestinal survey, but when question two asked if farting had affected my cognitive function, I decided it was possibly a bit detailed, and that if the diet reached that point I ought to stop. I settled on a simpler four-point survey covering flatulence, bloating, discomfort and borborygmi (which is an excellent onomatopoeic word for tummy rumbles). My first observation: it is quite tricky to eat the recommended amount of fibre. On a couple of evenings, I found myself desperately munching bran flakes to get across the line. The second observation was more surprising: I experienced no gut symptoms with the high-fibre/fermented food diet. I basically ate baked beans everyday – and as we all know, beans are good for your heart, but the more you eat, the more gas accumulates in your digestive tract. Based on this I expected seismic consequences – possibly even cognitive decline, but none of the things I measured changed: no bloating, no flatulence (at least no more than usual), no change in number or type of stool, not one borborygmus (the even more satisfying singular form). I can only speculate that the fermented food added in bacteria that broke down the fibre as it passed through – I could test this hypothesis by only eating fibre for a week without the fermented side dishes, but I fear that might test my marriage somewhat. So, with some relief, on the Friday of the test week, I stopped eating All-Bran and handed over a package of pestilence to one of Kinross's team of scientists in an unmarked bag underneath the statue of a man inexplicably looking in a shoe outside St Mary's hospital.

IT'S TIME TO PLAY ... 'WHO'S IN YOUR POO?'

A few weeks later, thanks to the stalwart effort of Dr Despoina Chrysostomou, I got the results back. What did I learn? Kinross proved to be correct: curry and beer had no effect on my gut microbes; poos 1 and 2 shared the same bacterial fingerprint. The bacterium Prevotella_9 dominated the top of the plops, with some contribution from *Bacteroides vulgatus* and *Faecalibacterium prausnitzii*. However, as you can see (Figure 19), I had a massive shift after a week on the brown stuff.

The Prevotella dramatically reduced in frequency and *B. vulgatus and F. prausnitzii* were usurped by *Bacteroides*

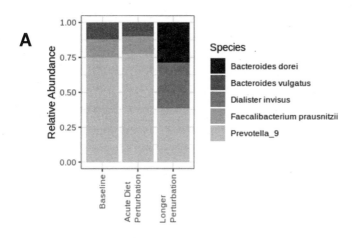

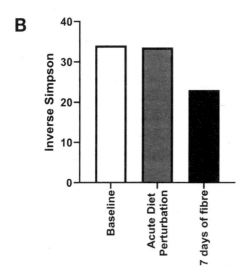

Figure 19. Changes in gut microbiome. Data generated sequencing my faecal microbiome taken over three timepoints – baseline (normal diet), acute diet perturbation (one day after curry and beer), longer perturbation (ten days after high fibre, high fermented food diet). Panel A shows top three strains at each timepoint, Panel B shows a measure of the overall diversity.

B. *dorei* and *Dialister invisus*. The commonly held view is that fibre increases bacterial diversity and so is a 'good thing'. However, my overall diversity had shrunk, not grown, after a high fibre diet. Researchers use different statistical ways to calculate diversity. Figure 19B presents one approach called the 'inverse Simpson index' in which bigger numbers mean more diversity, smaller numbers less. Bars 1 and 2 are greater than bar 3, indicating a reduction in diversity when I ate nothing but fibre. Essentially, I experienced a massive floral bloom of brand-new bacteria in my bottom. I investigated the scientific literature about my new top 3, assuming that since there had been a swelling in their ranks, they might be fighting on my side. Wrong again (maybe). One of the bacteria that increased, B. *dorei*, has been linked to type 1 diabetes onset in children;[28] one of the bacteria that decreased *F. prausnitzii* has been associated with good health outcomes.[29] I have no idea where any of my new bacterial buddies came from. Fermented foods are reportedly full of lactobacilli, but none of these appeared; the bacteria from the probiotic yoghurt I drank didn't touch the sides, disappearing without a trace, and the second-most frequent bacteria after intervention (*D. invisus*) normally lives in the mouth (and no, I didn't lick the poop).

Which shows us the following:

1. You *can* shift the gut microbiome, but you need to do something quite substantial over a sustained period.
2. If you put faith in a single bacteria in a swarming multitude of microbes being good for you then curry and beer are better for you than fibre.
3. We are all individuals; (at a bacterial level) what I see in my faeces will not be in yours.

4. The microbiome is an extremely complex system; don't waste your money getting your poop sequenced.
5. Conclusion 2 is not a sensible conclusion. Of course, curry and lager aren't good for you. Eat more fibre.

Ultimately, diet, microbiome and diabetes weave an intricate web that we are only beginning to unpick. But one thing for certain is that another food-linked risk factor directly correlates with the risk of type 2 diabetes – obesity.

WEIGHTY MATTERS

Type 2 diabetes is multifaceted, with genetic and ethnic risk factors. But at the heart of it, a key driver is being overweight. The simplest measure of obesity remains the very much maligned and misunderstood BMI (body mass index) devised by Adolphe Quetelet, a nineteenth-century Belgian statistician and sociologist. Quetelet founded the approach of using statistics to address societal questions, suggesting that any human measurement spreads over a normal distribution of values. To calculate BMI, take your weight and divide it by your height (squared). If you like formulas, $BMI=weight/height^2$; with results given in kg/m^2. If you use imperial measurements, I'd recommend calculating it online; but you can multiply lb/in^2 by 703 if you insist. Critically, BMI relates weight to height; weight alone isn't enough as a measure of obesity – being twelve stone (72 kg) is OK if you are six foot (182 cm) but less OK if you are five foot (152 cm). As Figure 20 shows, BMI falls in different bands, with anything less than 18.5 characterised as underweight, 18.5–25 normal, 25–30 overweight and 30+ obese.

Depending on the weighing scales I use, my BMI is either 23.4 (my parents) or 24.6 (the ones we found on holiday) – and since the bible says, 'Honour your father and your mother, so that you may live long,' I am obliged to choose the lower value. These values skirt just the right side of unhealthy; it's been a long time since I was a skinny teenager in the underweight bracket and listening to too much goth music. The BMI is imperfect, particularly around ethnic variance due to the dataset Quetelet used to set his original boundaries. He measured soldiers from the Scottish Highlands and French Gendarmerie; proving it is always easier to measure a captive audience than a representative one. And as a population measure, there are of course limitations – Dwayne 'the Rock' Johnson, wrestler, actor and part time singer, has a BMI of 30.7 (according to the internet at the time of writing) and there is no way I am going to call him obese, but 'You're Welcome' to if you want Dr Quetelet.

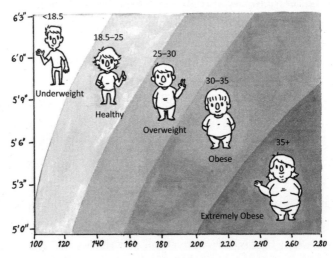

Figure 20. Body mass index (BMI). Shows how height and weight are related in BMI.

Whichever measurement system used, obesity can definitely be linked to type 2 diabetes. Any number of studies have shown this – for example one meta-analysis estimates that being overweight increases the risk of type 2 diabetes threefold; being obese increases the risk eightfold.[30] I am not demonising obesity; the identity politics of weight fall well beyond my remit. Body shape and size are as much personal choices as smoking and drinking; but all of them carry risk, especially when combined.

What can be done? Basically, eat better. But what does better mean? At this I will pause and take a deep breath before walking into the diet minefield. To help me disentangle good from bad, I spoke to Professor Gary Frost, a dietician and Chair in Nutrition and Diabetics at Imperial. I asked him how diet affects ageing, to which he replied, 'I think that's a bit of an open question at the moment. What we do know is that overconsumption is problematic. If you become obese or overweight, then that affects your lifespan and your number of disease-free years.' The killer question is how to avoid becoming overweight?

Ancel Keys was one of a long line of experts, semi-experts and downright quacks proposing diets for longevity and weight loss. Diet books shift huge numbers of copies. The Atkins diet book sold more than a million copies in the UK; remarkably, 'Dr' Gillian McKeith's poop-sniffing *You Are What You Eat* sold even more – and she doesn't even have an accredited PhD!* All of these books suggest subtly different ways to be fitter, happier, more productive. Much of the advice about removing specific things – cutting carbs, cutting

---

* Yes, I am massively jealous of the sales, and no, having a real PhD doesn't really make up for that.

fruit, cutting dairy miss the point because they don't look at diet in the round. Luckily for me I work with colleagues who can simplify the more complicated message – back to Oliver: 'You can't outrun a bad diet,' he explained. 'There are no bad foods, there's only bad diets.' He gave me an example – his son loves donor kebabs, which sit in no one's list of healthy foods, and occasionally they will partake in one for supper (presumably his son didn't consume the prerequisite ten pints of super-strong lager beforehand, I thought it impolite to enquire). One donor does not a bad diet make; eating unhealthily, occasionally, in the context of an otherwise healthy diet does not make you unhealthy. Another way of looking at diet is energy density – calories per amount of food consumed. Frost again: 'Some diets are more protective – those that avoid energy-dense foods and saturated fats.' That said, the 'quality/cleanliness/naturalness' of the food is not a critical factor; both Frost and Oliver told me that the idea of ultra-processed foods (UPFs) was not straightforward, because they lack a tight definition and because there is so much overlap in the Venn diagram of high-fat foods, high-salt foods, high-sugar foods and UPFs, making separating the bad component(s) out tricky.

The underlying message in the context of weight and diabetes is that your calorific intake should meet your calorific need; ideally the calories should not come from energy-dense foods (you can't just eat chocolate). This advice can be further refined to say: reduce red meat (cancer), salt (blood pressure), saturated fat (heart disease) and increase fruits (for antioxidants and other goodies) and fibre (to feed the microbiome).

Of course, 'have a good balanced diet' is much easier to write than do. We have such easy access to calories; the three chocolate fingers I just ate (because we ran out of caramel wafers) contained 108 calories, apparently enough to sustain one hour of writing, so I should resist the urge to head back downstairs for more snacks. But they're so tasty – and recent research demonstrated that high-fat/ high-sugar snacks alter our reward circuitry; essentially, they are a bit addictive.[31] The ease of calorie acquisition combines with reduced physical output to ill effect; on average people's jobs require fewer calories than one hundred years ago, because fewer people perform manual labour. Furthermore, we burn fewer calories as we age. All of which means calorie intake exceeds energy expenditure and therefore weight increases, and with it risk of a range of diseases. Cheap, high-calorie food, with its moreish quality, makes dietary restriction challenging. What else can be done?

BEYOND DIETS

'Nothing tastes as good as skinny feels'

Kate Moss

When diet alone fails to control obesity and prevent or reduce diabetes, what other options exist? The hope is for a miracle pill. The hunt for a silver bullet has a long and often nefarious history. For the Victorian tapeworm diet, people would consume parasites in order to lose weight; some dark corners of the internet still suggest this as a good idea. It isn't. Dr Ziyin Wang in my lab tested the common observation that the immune response to infection suppresses the appetite; it does, but having a perma-cold doesn't seem sensible.[32] Following their use to keep soldiers awake in the war, from the 1950s to the 1970s doctors prescribed amphetamines as weight loss drugs. Speed's even shadier cousin meth(amphetamine) has also been used in diet pills. Though the side effects of psychosis, hallucinations and death nipped that one in the bud. An alternative approach, orlistat (Xenical), prevents fat reabsorption; whilst highly effective the fat from food must go somewhere – in this case it comes out as steatorrhea, oily foul-smelling poo, often without warning. Which frankly puts it alongside the side effects of the Meth-diet™ (and sugar-free gummy bears).

This all leads us to the current miracle drug *du jour*: semaglutide (brand names Ozempic/Wegovy). The drug mimics the action of glucagon-like peptide 1 (GLP1) which in turn comes from the same protein precursor as glucagon – the anti-insulin. But instead of causing glucose release, GLP1 leads to the release of insulin, reducing both blood sugar and appetite. Whilst it has made a large impact on the cultural consciousness, semaglutide is not the first GLP1 mimic. One of the first drugs in this class was exenatide, which, like ACE inhibitors, first came from reptile venom – specifically the Gila monster. GLP1 was discovered by Svetlana Mojsov, a Yugoslavian-born chemist, whilst working in Boston in the 1980s, and GLP1 mimics have been part of the endocrinologist's tools to manage T2D for twenty years. Semaglutide causes fewer side effects than previous iterations, making adherence easier. The semaglutide bubble expanded via A-list actors and TikTok influencers to an extraordinary level of demand, which, in 2023, failed to be met by supply. But the weight loss achieved is rapidly reversible; once you stop taking the drug, appetite returns and potentially the weight too. Time will tell as to the long-term impact of semaglutide and the other copycat drugs that inevitably will be developed. One of the more striking claims is that semaglutide could increase airline profitability by reducing the average weight of passengers but affect the shares of junk food manufacturers such as Pringles, Jell-O or Fritos. The economics of it blow the mind; in 2023, semaglutide's producer Novo Nordisk was worth more than the rest of the Danish economy put together, selling so much drug it prevented a Danish recession in Q2; though presumably it affected beer and bacon sales.

But if semaglutide is too pricey, another much more widely available compound might actually be a miracle drug: metformin. Metformin was originally isolated from *Galega officinalis* which amongst its other common names has been called 'professor weed' – who probably teaches philosophy at an art college near you. Metformin was first shown to reduce blood sugar in the 1920s but remained ignored because of other more shiny compounds – such as insulin. A French doctor, Jean Sterne, rediscovered its role in 1957.[33] Subsequent studies showed that metformin use significantly reduced cardiovascular risk and can reduce the incidence of diabetes. This led to metformin being the most prescribed glucose-lowering drug, placing it on the WHO list of essential medicines. But the ability to reduce glucose is not the only reason it is a wonder drug: like aspirin it provides a range of other benefits, but unlike aspirin we don't really know how it works. Cohort studies also suggest it can reduce the risk of cancer and dementia. In our lab we have shown that by reducing lung glucose, metformin can reduce the severity of respiratory infection; admittedly we used fat mice.[34] We extended these studies into humans with COPD, working with Drs Fiorenzo and Farne, my erstwhile lung advisors from previous chapters. We didn't see any effect of metformin in reducing lung glucose; this was particularly sad because of how difficult the study had been to set up and run, making me shudder with every email. Metformin might even slow ageing, but this might be over-stated – the best evidence comes from tiny nematode worms (C. *elegans*), and my mate Cathy (Dr Slack) showed this doesn't even translate to fruit flies,[35] let alone humans.

Drugs for weight reduction, particularly short-acting ones, are effective only as long as they are taken. Other approaches grouped under the heading 'bariatric surgery' can provide longer-term solutions. Three common opera-tions are performed: gastric sleeve, gastric bypass and gastric band. Fundamentally, they work by reducing the size of the stomach and therefore the amount of food that can be eaten. This type of surgery is recommended in the UK for people with a BMI over forty. It has mixed outcomes – according to one meta-analysis it can significantly reduce a range of diseases associated with obesity, including diabe-tes, cancer and cardiovascular disease. But it carries an increased risk of suicide and self-harm.[36] The reasons for this are poorly understood, but it should be noted that bari-atric surgery does not happen in isolation; underlying psychiatric conditions may also contribute. There are complex behavioural issues around food that will not neces-sarily be fixed by surgically curtailing the overeating.[37]

LIVE, FAST, DIE OLD

Bariatric surgery effectively works by reducing the amount eaten, and if surgery sounds a bit irreversible, another approach has gained a lot of attention in the last ten years: fasting – specifically, intermittent fasting. This can take various forms, for example the 5:2 (five days eating, two days fasting) or the 16:8 (sixteen hours fasting, eight hours eating). The underlying principle is to not eat for extended time periods. One of the more ludicrous versions is the 'warrior diet', whose name alone screams toxic masculinity. It suggests fasting for twenty hours and then stuffing yourself for four hours at night – like 'warriors of old'.[*] The four-hour food intake period comes with no restrictions on food type – it is not a good diet plan. Other intermittent fasting plans have a bit more underlying science and can support weight control. Intermittent fasting differs from calorie restriction (just eating less); but only subtly, the main perceptible difference is in timing.[38] Calorie restriction diets alter the total intake; intermittent fasting focuses those restricted calories into specific time periods. If you eat one thousand more calories than you expend but in an eight-hour window, you will put on weight. One possible benefit of intermittent fasting comes through adherence: for some people it can be easier to focus willpower on a limited number of days and then eat more freely on others.

Reducing calories, by whatever method, can improve and even reverse type 2 diabetes. But as well as reducing weight and the impact of diabetes, calorie restriction, particularly through fasting, may provide an additional benefit – extended lifespan. This has been observed in the lab. For reasons of practicality, most lifespan studies are performed on lab animals – specifically nematode worms, fruit flies and mice, all of which live short lives and can be studied intergenerationally within a single three-year PhD. This leads to challenges in generalising results to humans

---

[*] I thought it might help to say 'warriors of old' in a terrible Viking accent to give it a fig leaf of credibility. It didn't.

– as Oliver put it: 'I'm not a fruit fly, and neither are you.' That said, reducing calories in all of these species can extend life, suggesting a common mechanism conserved across the animal kingdom. A lot of hyperbolic journalism has been penned on this subject, which nearly always makes a reference to how our palaeolithic ancestors would have endured periods of enforced hunger when the mastodon hunt failed and we evolved to benefit from fasting. This makes a fairly large assumption that we basically stopped evolving once we left the cave, which may ring true for Millwall FC supporters, but evolution for the rest of us has carried on ticking over.

Cynicism aside, one of the alluring aspects of fasting is that it can trigger a cellular pathway called autophagy (Greek for 'self-eating'), where cells recycle material from within themselves. This loops back to the ideas about ageing shared with me by Slack and O'Neill: as they senesce, cells become damaged by their own contents. The drawings of cells we make at school (basically a fried egg, with the nucleus as the yolk) are simplistic to the level of being misleading. Strings of proteins criss-cross the cell, with subdomains based on biochemical gradients and driven by localised pH. A lot of stuff happens in each and every one. Over time, the cellular proteins begin to degrade and in so doing stop interacting with the things they should and potentially start coupling with other chemicals; which in turn can affect the function. Likewise, the mitochondria, those cellular powerhouses, begin to revolt as we age, chucking out their DNA and stressing out their cellular landlords – DNA in the wrong place being a major cellular red flag. Autophagy acts as intracellular housekeeping, removing old proteins and breaking them back down into their component parts, so other ones can be built anew. This process kicks into gear during sleep and may in fact be a primary reason why we need sleep. All of which means that just a bit of starvation, synchronised with the time when the body is at rest, could be beneficial.

Whether uniquely beneficial, or simply as a way to cut calories and reduce obesity, fasting passed my self-experiment threshold. However, having spent a long time deciding whether or not to fast, I then picked the first diet plan I saw on the internet – I'll admit the shiny *Science* paper linked to it lured me in.[39] Hence, for one week in September 2023 and quite without the support of my wife or family or work colleagues,* I started a five-day 'fasting-mimicking' diet. Although calories can be cheap, a premium calorie-restrictive diet can be incredibly expensive. The diet I chose gave me less than nine hundred calories per day, and the five days' worth of food cost a staggering £160 – for the same amount I can buy eighty packs of caramel wafers, providing some eighty thousand calories. I felt so sickened by this I immediately gave the same amount of money to a food bank. The whole situation reminded me of the character Famine in Terry Pratchett and Neil Gaiman's brilliant *Good Omens*, who had successfully engineered a food with a nutritional content 'roughly equivalent to that of a Sony Walkman'.[40] Anyway, I found myself in receipt of five 'delicious' days of diet plans. As a condition of doing the diet, I had to book the week into the family diary, so everyone knew to avoid me. I inadvertently booked it into my wife's calendar too so everyone at her work knew and also avoided me. In

---

* Genuinely no one I know thought this a good idea, except my mate Alan – who doesn't have to live or deal with me on a daily basis.

addition, I warned my lab team who would have to endure my self-inflicted hanger.

In a vague attempt to be scientific I recorded as much information as I could about my body and, more importantly (for those around me), mood in the weeks before during and after the fast. I took inspiration from one of my favourite papers of the year 'Veganism and body weight: An N of 1 self-experiment';[41] which contains the immortal line 'Self-experimentation and N of 1 studies are underused but potentially valuable research tools'.* The diet started looming large on my consciousness from about a week out. To such an extent that like a squirrel I started hoarding calories for the long winter ahead – leading to me eat more unhealthily, consuming doughnuts, duck vindaloo and lager in a last supper of excess.

The whole week's food came in a very shiny package, with individual boxes for each day. Each box was not very big and mostly empty, and the scant contents not very enticing. My meal plan for the week comprised: powdered soup (ten sachets) supplemented by five nut bars, some suspicious kale biscuits (three packets), a handful of olives (five packs of eight olives) and three 'chocolate' bars made of chicory root fibre (i.e. not chocolate). This was all supplemented by some herbal tea and a suspicious drink supplement whose main purpose was to keep my guts regular in spite of the near total absence of solid food. If you want an easy way to imagine my experience, think of Eric Carle's *The Very Hungry Caterpillar*. Each day I ate two soups, until Saturday when I ate a full English breakfast, a huge plate of pasta, a

---

* This tickled me because I spend a lot of my life explaining to students why N of 1 experiments are *not* valuable research tools.

curry and drank four pints of lager. The question being: did I turn into a beautiful butterfly?

I would say the first day was the worst, but the third and fifth were also pretty tricky. I finished my nut bar breakfast at 7:30 on day one and was hungry by 8:15, not a great start; same at lunch – eaten by 12:15, hungry by 12:45. Time slowed during the week. One of my notes from day one simply says, 'I'd like a snack now'. I found it marginally easier on day two – but happiness is relative, the extra olive in that day's package being a cause for joy. By Wednesday I started hoarding crackers from myself, and then it started raining; this was the nadir. I attempted some different ways to score happiness (the UK government's ONS4 question-naire[42]), depression (the PHQ-9[43]) and anxiety (the GAD7[44]), and I tried to make it more objective by hiding the previous day's results from myself. As can be seen in Figure 21 – my happiness dipped, and depression peaked on the third day of the diet (wet Wednesday). Once I passed the hump, things improved, though hunger pangs weren't helped by one of my colleagues having smoked salmon salad for lunch and my son having a mountain of curry for tea in front of me – whilst I sipped a small bowl of soup. Admittedly by this point I had gone wild and started adding pepper, cayenne and the occasional basil leaf to the soup. The week did finally end – with one last soup.

What did I learn? I had absolutely approached this with a closed mind. I fully expected to hate it, and quite worried about what it would do to my relationships with colleagues and family. My anxiety reflects this spiking the day before the diet – how would going low-cal affect me? In terms of general outlook, it was much better than expected. One of my team said on Friday, 'You've not been grumpy; in fact, you were *less* grumpy than usual!' My wife thought I just had too little energy to make a nuisance of myself.

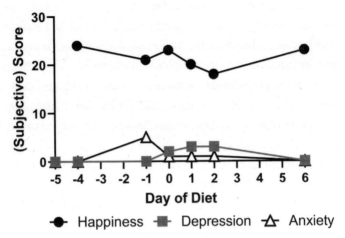

Figure 21. **How I felt during my diet.** Entirely subjective scoring system to assess how the diet affected my mind (using established questionnaires).

I found compliance to the diet tricky, especially when at home, with all kinds of tasty treats in view, but not impossible. It certainly wasn't pleasant. Not only did I cut out food, I cut out drink, reduced exercise and opted out of my social life. I discovered food plays a pretty central role in my life. Less eat to live but more live to eat – I spend a lot of time thinking about food, preparing food, eating food. Not eating gave me more time, but nothing with which to fill it because I had no social life or energy to exercise. I spent two heady nights filling out my tax return. It was no way to thrive long-term. I also became much more aware of the omnipresence of food – every third sign advertises food, every other shop is some kind of food outlet, the Circle Line interchange at Paddington smells of fried chicken and at

Victoria it smells of steak. Another downside was the lack of fresh fruit and anything with any substance to it. I do wonder what havoc it caused my microbiome, and I broke my fast with an apple fresh from the tree in the garden; it was perfect, the very model of an English autumn, dewy and as crisp as the surrounding air. I also discovered doing an extreme diet can be added to the list of things that you must tell everyone you meet (alongside veganism, cycling and Oxbridge – see Chapter 7). It dominated conversations for the whole week.

But these downs were countered by the loss of half a stone (three kilograms) in a week, a significant drop in waist size from ninety-four centimetres (the top end of low risk) to ninety-two centimetres and being able to fit back into a pair of army trousers that I have owned since I was a whippet-thin eighteen-year-old officer cadet. My BMI also fell from 24.2 (on the edge of bad) to 23.3. The challenge remained of whether I can keep the weight off and stick to my fresh resolve to snack less, eat only at mealtimes and avoid excesses of calorie-dense foods. I weighed in a week later and found I had kept it off, but a couple of months later, after a week's conference in Belfast where the Guinness flowed free, my wife commented that maybe my tummy had returned. And by the time I came to read the final draft six months later, there was no denying middle age had resumed its spread. Still, for a month in September I was slim and beautiful!

As I said, I went into the diet with a determination that it would be a terrible and a fundamental waste of time. It wasn't. But one consequence of the diet – basically soup and flavoured water for a week – was that I found myself endlessly peeing. And so, we head downstairs to the final

body system: the liver and kidneys. I have kicked these along
the road, hoping that by journeying through other tissues I
may have found the willpower, or maybe a workaround, to
deal with the impact of my own personal health *bête noir*,
even worse than chocolates or crisps – the demon drink.

CHAPTER 10

Plumbing New Depths: Killers Number 7 and Number 8, Liver and Kidneys

'Therefore, much drink may be said to be an equivocator with lechery. It makes him, and it mars him.'

William Shakespeare, *Macbeth*

On Remembrance Sunday, whilst standing in a lengthy queue of (predominantly male) veterans on Horse Guards Parade waiting for the gents, one stereotype of ageing was writ large: the increasing need to pee. In recent years, my mental map of London has become landmarked by public toilets rather than bars or nightclubs. I am not the only one; eighty-one percent of those surveyed by Age UK found toilet provision in the capital to be inadequate. London provides only 1,500 public toilets – working out as one per five thousand people. Admittedly there are pop-up loos, street corners and what the *Viz* comic describes as the McSh*t – going into a McDonald's with no intention of treating yourself to a Happy Meal. But the situation is far from satisfactory and can cause health consequences; fifty percent of those surveyed admitting to restricting their liquid intake before going out, increasing the risk of dehydration and falls.[1] It can also lead to legal consequences – a man was recently fined for littering when caught urinating in a lay-by: just in case you wondered, the rumour that you can legally pee on your own rear tyre is of course an urban myth.

On average we pee six or seven times a day. Lots of factors affect the rate we micturate: some volume-intrinsic – if you drink more water, you wee more; some drug-induced – both caffeine and alcohol increase urgency; and some medically influenced – urination frequency (and sweetness) being a hallmark of diabetes. Three organs enable the production and release of urine: the liver, kidneys and bladder (with some input from the brain). As we age, our kidneys get less efficient at recovering water, increasing the flow to the bladder, which in turn loses elasticity, reducing the volume it can hold.

Peeing does two things: it clears toxins and removes excess liquid. The reason we need to get rid of toxins should be fairly obvious, because they are toxic! The liver and kidneys play a central role in this process. Obviously not ingesting poison in the first place reduces their workload, but we cannot avoid some toxins: cellular breakdown products can be damaging, as are the various biproducts of the bacteria that live in us. And of course, a huge number of us deliberately and frequently ingest one particular poison – alcohol.

Why we need to control water is a bit more nuanced. The correct term for water control is homeostatic osmoregulation. Homeostasis is maintenance of bodily systems at a constant(ish) level; osmosis is the movement of water up a solute gradient.* You may remember osmosis from school

* This is most often portrayed as being about salt, but all sorts of things in your blood affect its concentration, including protein, and not forgetting glucose which is why people with diabetes need to pee more.

– putting a bean in a salty solution to see it shrivel and then returning it to water where it swells. The liquid part of the blood, plasma, needs to be maintained at roughly the same overall concentration of salts, sugars and proteins to stop anything it touches shrivelling or swelling up. Red blood cells are particularly sensitive to changes in salt concentration; if the plasma becomes too dilute, they explode.

The body controls water volume via a circuit between the kidneys and the brain. Surprisingly, it doesn't involve the medulla, but it does involve two other cerebral stalwarts – the pituitary gland and the hypothalamus. The pituitary looks like a pair of tonsils, dangling off the hypothalamus. The hypothalamus receives electrical signals, and the pituitary converts those signals into hormones that then act on the rest of the body. Pituitary hormones influence many bodily functions, including growth, lactation and sex differentiation. To control urination, the pituitary releases a hormone called vasopressin or ADH (antidiuretic hormone) on the instruction of the hypothalamus, which constantly scans the concentration of the plasma (as seen in Chapter 5, salt levels play a key role here). If you have too little liquid in your system, the concentration goes up; if you have too much liquid it goes down. Increases in concentration (indicating too little water) lead to thirst (making you drink more) and the release of ADH. In turn ADH increases the amount of water retained by the kidneys.

The kidneys act as a filtration unit for the blood. Each time your blood passes through them, the liquid part gets squirted out of the blood vessels into the tissue of the kidneys; this happens at a sieve-like node called the glomerulus. Lots of things are dissolved in the plasma, and they leave the blood vessels at the same time. The colour and smell of the urine depends a lot on what is or isn't filtered through the kidneys. Taking too much vitamin B_2 turns your urine bright yellow – a side effect that can be quite alarming if you have just taken a fizzy vitamin pill to fight off a hangover. And of course there is beetroot and its terrifying effect the first time you eat it to excess – the pink urine has its own name, beeturia. Surprisingly, only fifteen percent of people get this – I had assumed it was universal, but I guess it's not necessarily something one discusses outside family. There is a military urban myth that drinking a glow stick makes your pee glow – but a whole level of stupid prevents me from trying or recommending that. If you *really* need glow-in-the-dark piddle, apparently tonic contains enough quinine to make it fluoresce under UV light.

Smelly chemicals also pass through; Sugar Puffs make my urine smell sweet, which apparently is unusual and maybe a sign of diabetes. Diabetes, by increasing the amount of sugar in the blood, overloads the kidneys, leading to more saccharine urine. But as my tests showed, my glucose control is pretty normal, and I've noticed the malodorous effect of the Honey Monster since childhood, so I'm not too worried about it. If someone wants to fund a study about it, I'd be very supportive; I'd even volunteer my wee (for science). Use of that most scientific tool (the online social media survey) showed that at least four other people out there have the same issue (from a survey size of fourteen, and six of those asked responded 'seriously stop asking'). The most famous urinary odorant is asparagus – whose distinctive aroma was remarked upon by Proust and

Benjamin Franklin.* Whilst everyone makes the stinky sulphur compounds (S-methyl thioesters if you want to synthesise your own), most people can't smell them (nearly sixty percent according to one study). The authors start their study with the excellent self-deprecating line, 'Few scientists have sought to examine the inherited factors associated with asparagus anosmia.'[2] Asparagus anosmia is associated with 871 SNPs, located on chromosome 1 and mostly found in two genes, OR2M7 and OR2L3; the functions of these genes remain unknown. The authors conclude that 'the root of human olfaction is not fully understood'. My own gene testing tells me I am able to smell it – as does my nose.

But the kidneys don't just filter things out, they need to recover liquid, and this is where ADH comes in. Through a remarkably clever piece of plumbing called countercurrent multiplication, the excreted liquid flows back into the blood, leaving the guff behind. This all happens in the loop of Henle, discovered by Friedrich Gustav Jakob Henle who was not sparing with eponymising body parts; he also coined the crypts, glands and tubes of Henle, Henle's fissure, Henle's ampulla, Henle's ligament and Henle's spline. Basically, he cut apart dead bodies and named stuff after himself, and, frankly, why wouldn't you? The level of ADH in the body determines the amount of water reabsorbed into the kidney (Figure 22).

---

* Demonstrating that my childish obsession with bodily function shares high-minded company. Franklin went further, noting that eating a small amount of turpentine made his piss smell of violets – not something I'd try; probably he rewired his brain when kite-flying.

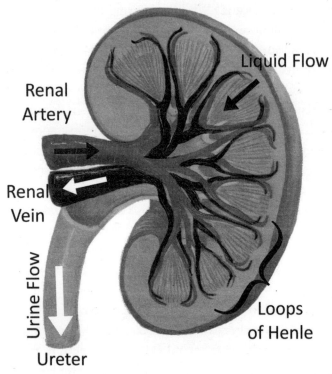

Renal
Artery

Liquid Flow

Renal
Vein

Urine Flow

Loops
of Henle

Ureter

Figure 22. The kidney is very serious business. Filtering our bodily fluids forty times a day, the kidneys do a lot of behind-the-scenes housekeeping to keep us alive. They also make urine.

Dehydration signals to the pituitary to release ADH, but its antidiuretic function can be overridden by other chemicals. The most obvious of which is booze. As Macbeth's porter acknowledges, drink is a provoker of nose-painting, sleep and urine. Net diuresis volume results from a combination of alcohol strength and liquid volume ingested: eight pints of ace super strong lager will make you pee more than a small glass of sherry. Extraordinarily, it took until the 1940s to link peeing to the ethanol component. Grace

Eggleton, working at UCL and the Maudsley mental hospital, measured the amount of urine produced relative to alcohol intake in a single subject over multiple days, showing a direct linear relationship.[3] The starting hydration state also has an impact – if you are really dehydrated, drinking beer will not make you pee. In another important study from the 1980s, small beer (less than two percent alcohol) restores hydration after exercise but anything over four percent slows rehydration down, which means that, in spite of the efforts to the contrary of club rugby players everywhere, lager sadly doesn't count as a sports drink.[4] However, another study suggests that adding sodium to light beer improves fluid replacement;[5] so if you drink beer after sport, eat crisps too – which, even as I write it, I can see is terrible advice. It needs to be noted that 1980s alcohol research was pretty wild; in one letter to the *Lancet* the researchers measured the response of a subject who drank sixteen pints of beer (kindly bought for him by the first author) in twenty minutes 'quite effortlessly' (admittedly in Newcastle-upon-Tyne); I'm still not quite sure why or what it shows.[6] What happened after the sixteen pints is unclear; it feels quite likely that the volunteer hit the heads. And despite its penetrance into popular culture, the beer seal (the notion that once you start peeing at the pub, you will continue with increased frequency) does not exist. The best explanation for 'breaking the seal' is a lag between fluid intake and release, and as most of us are probably slightly dehydrated at the beginning of a 'session', it takes some time for the liquid to reach the bladder. But once we get going, it is a one-in one-out situation – the bladder being relatively small – only holding about half a pint and feeling full when it contains a quarter of a pint.[*] Drinking one pint of beer therefore will fill that stretchy bag of urine. And holding it in increases the risk of UTI and long-term damage. If you need to go, go.

The other commonly consumed diuretic is caffeine, which like most recreational drugs is pleiotropic – having a wide range of actions. Broadly speaking, caffeine (or 3,7-Dihydro-1,3,7-trimethyl-1H-purine-2,6-dione) is a stimulant. It can be lethal, but it would take 10 g to kill you – the average cup of tea only contains 50 mg, meaning you would probably drown before overdosing on tea. Caffeine is mildly addictive – withdrawal causes headaches. Caffeine works by binding to adenosine receptors, which are linked to both tiredness and water retention by the kidney, explaining why tea makes you pee.

Other external factors influence the need to urinate, particularly temperature. The cold causes our blood vessels to contract (to keep the warmth inside), increasing blood pressure, forcing more liquid through the kidneys and therefore out into the bladder. Factors often work in tandem. One of the curious quirks of British military etiquette is not being able to pee before the end of a meal, or before the Royal Sovereign leaves the building (or at least line of sight). As I learnt to my discomfort when I went to watch my friend's daughter on parade in front of King Charles III. It was an early start and very cold, so a cup of tea seemed sensible at the beginning of the two-hour parade. By the

---

[*] Whilst I have access to measuring cylinders at work and home, it felt poor form to use them for this purpose. It took me longer than it should have to think of a way to measure my own urinary volume. I finally realised I could just use a milk bottle!

end, less so. The ultimate kicker was that the cadet we had come to see had broken herself somewhere on Dartmoor and was thus reduced to driving dignitaries around in a golf cart. Still, we got to see the back of the new king's head from three hundred metres away – well worth a vast amount of discomfort, probably.

Of course, you don't even need drugs to change the amount you pee, you can just drink more (or less) water. Standard advice says to drink six to eight cups (about two litres, or four pints) a day. This liquid can come in any form, but tea and alcohol might be a net negative because of their diuretic effect. Being correctly hydrated is no doubt important, and one of the first ways you might notice dehydration is through headaches. Whether dehydration directly causes headaches is not clear.[7] There seems to be a generational relationship between water and hydration; young people these days definitely consume more water than their parents or grandparents. My limited daily liquid intake is small enough to make a millennial run for their enormous water bottle – on a standard day it includes the milk in my cereals, two cups of tea (mid-morning), two cups of water (lunch), a fruity tea (if am feeling fruity) in the afternoon, a glass or two of water with dinner and maybe a beer. This crosses the NHS threshold, but not the 'wellness' advice from younger people and 'health influencers' who want us to drink so much water that we pee constantly. The good news for those of us who want to choose studies selectively that support our view is that a 2022 paper in *Science* reassessed water requirements, suggesting a catch-all of two litres per day may not be right or even supported by evidence.[8] One volume fits all doesn't work for water because a lot of liquid comes from food, and water requirements reflect a wide range of variables such as size, exercise and temperature. What's most satisfying about the report from a generational-war point of view is that you can argue that excess water intake damages the environment because of the resources to purify it on the way in and out. Even better, another study found a wide range of microplastics in plastic bottles,[9] making tea drinking safer for you anyway.

KIDNEY FAILURE

Your kidneys work tirelessly to cleanse your body; with every heartbeat the blood squirts through the nephrons, removing all of the bad and smelly things; they filter 180 litres a day (given our bodies contain five litres of blood, that means each litre gets cleaned at least thirty times a day). The faster your heart beats, the more the kidneys filter – hence we need to pee when we are nervous; adrenaline increases circulation, pushing more fluid through the system. It's not overly surprising that kidneys fail! There are 7.2 million people in the UK with some form of chronic kidney disease – that's one in ten. Three million of them have symptoms, seventy thousand have the most severe form and forty thousand die of kidney failure each year. The kidneys can also fail acutely. I spoke to David Thomas, professor of renal medicine at the University of Cambridge, about the causes of kidney failure. If you have been paying attention to this point you will probably be able to guess the causes: diabetes, heart disease and elevated blood pressure all affect kidney function. Diabetes damages the kidneys in the same way it does other peripheral vasculature – by increasing the pressure and screwing up the plumbing. I asked Thomas whether he had any good anecdotes about the kidneys to which, flicking back his fabulous mane of hair, he replied in a strong Welsh accent, 'The kidney is a very serious business, John.'

Kidney failure occurs when they can no longer remove waste or liquid from the body, leading to fluid retention, swelling, cramps and tiredness. Mild cases of chronic kidney disease can be managed through lifestyle change and medication. But more severe kidney failure may necessitate transplantation. Surgeons could swap out kaput kidneys before hapless hearts – because you can survive longer without a kidney than a heart, but not by much. The first, unsuccessful, human-to-human kidney transplant was performed in 1933 by Soviet surgeon Yu Yu Voronoy, followed by a string of equally unsuccessful studies in France in the 1940s using guillotine-fresh kidneys.* But freshness hadn't been the problem; whilst the Soviet kidney might have necrosed having been removed six hours after death, *le rein Français* was execution-fresh. The problem, as with all transplants, was the immune system.

Peter Medawar performed much of the underpinning work in transplant immunology. As with many eminent British scientists, Medawar's parents were immigrants, and he was born in Brazil – where his Lebanese father worked as a toothbrush salesman. During the Second World War, he worked in a skin transplant unit, trying to repair burned airmen, where he noticed a more rapid rejection of grafts donated from other people. He continued this work after the war by moving skin between twin cows. In 1953 he

---

* Madame Guillotine had a long career, starting in 1792; she only stopped claiming victims in 1977.

demonstrated that rejection could be overridden but only if cellular transfer occurred in the womb, indicating an immune component. Medawar's idea of actively acquired tolerance formed the basis of his Nobel Prize. More importantly, his ideas that the immune system played a role in graft rejection and it could be trained to accept tissue from someone else formed the basis of transplant surgery.[10] Researchers explored a range of approaches to prevent the immune response targeting the donor organ, including irradiation and nitrogen mustard, the toxic forerunner of cancer chemotherapy. By the 1960s it was possible to transplant a kidney, but the survival rate remained extremely poor – ninety percent of recipients died. The breakthrough came at a conference in 1963, when a surgeon (Tom Starzl) reported a seventy percent survival rate, to which the audience reacted as scientists do when challenged – by pooh-poohing his research. Fatal error! He had added a different type of immunosuppressive drug (the steroid prednisolone) with dazzling effects. Suppressing the immune system with drugs transformed the success of transplant operations; two of the main drugs used being tacrolimus and cyclosporin. Both drugs were discovered in the quest for novel antibiotics, using the same approach used to find statins (sifting through mud). Microbes are the apothecary of modern medicine. Tacrolimus comes from a soil bacterium, cyclosporin from a fungus. Both drugs turn off T cells in the person receiving the foreign organ, preventing rejection. But the immunosuppression needed to accept the donor organ also exposes transplant recipients to a particularly high risk of infections.

Whilst drug suppression of the immune system improves transplant acceptance, the final piece of the puzzle comes back to Medawar's studies transplanting the spots between twin cows. It was clear that family proximity determined the rejection or not of a transplanted organ. Transplants between identical twins were more successful than non-identical twins, which in turn were more successful than between siblings, which far outstripped getting a kidney from a stranger – even if fresh from the guillotine block. This comes back to the way in which genes are shared between generations – you get half of your genes from your mother and half from your father, your siblings also get half from each parent and because of statistics end up with about half of the same genes as you. In the context of transplant immunity, the human leukocyte antigen genes (HLA) play a key role. They encode a cluster of proteins that helps immune cells identify whether the tissue is self or foreign and hence form the basis of tissue rejection; they evolved to help our cells spot infections. HLA genetics are extremely complex – the gene family is spread over six gene loci, and each locus can contain one of thousands of possible variants, making the likelihood of an exact match extremely rare. This contributes to the huge challenge of matching organ donors and means it pays to stay on good terms with your siblings – they might be carrying the kidney you need in the future! Remarkably, in 2021/22, surgeons transplanted nearly three thousand kidneys in the UK, and advances in tissue matching and immunosuppression mean a transferred kidney can last up to twenty-five years in the recipient. As with other organs, the future might include xenotransplantation (from pigs) or kidneys grown in a dish, but both are a way off. The kidney is after all *serious business* and artificially growing one involves a lot of plumbing.

Despite advances in kidney transplantation, a long waiting list remains, with five thousand people currently awaiting a kidney in the UK. Luckily, they can be kept alive whilst they wait – through dialysis. This machine does the job of the kidneys. It fires the blood through a semi-permeable membrane which, like the kidneys, dumps out the bad and keeps the good. The first dialysis machine was built by a Dutch doctor called Willem Kolff, who developed it whilst the Netherlands were under German occupation in the Second World War. Animal research indicated that cellophane could act as a membrane to filter out bad stuff. Conveniently, sausage casings are made from cellophane, and Kolff used twenty metres of sausage casing,* connected via orange juice tins to an old washing machine drum that spun it through liquid at a different concentration causing the bad chemicals to flow out. The first fifteen attempts were a busted flush, but eventually, through adding anti-coagulant, Kolff twisted successfully, and Maria Schafstad (a known Nazi collaborator) regained consciousness after eleven hours on the Heath Robinson dialyser. The development of dialysers continued to have a 'dad built this in his shed' feel right up to the beginning of the twenty-first century, but they now form a central part of kidney treatment – albeit time-consuming and disruptive, with multiple sessions a week of up to four hours each.

BLADDER-WRACKED

The excess liquid that leaves the kidney drains out into the bladder. It is a bit of a curious organ; basically a collapsible

---

* Virchow would have been proud.

bag of wee that sits inside your hips. Different species evolved different mechanisms to carry their urine around with them. Birds have no bladder at all, basically leaking a stream of urine across the skies.* At the other extreme, the *Bufo* toad can hold up to half of its body weight as water in its bladder. Why bladders evolved is not entirely clear, the most likely explanation is as a way to escape predation; having a stinky trail leading to your nest probably isn't a great survival strategy. The bladder could also act as a defence mechanism – pick up a puppy and the stream of golden excitement might be sufficiently surprising to make you drop it again. The muscles at the neck of the bladder and the top of the urethra control the flow of urine; when you pee, the bladder contracts and the urethra opens. As we age, bladder elasticity decreases and therefore urination frequency increases – since it can hold less. Peeing frequency can also be affected by enlargement of the prostate blocking the tubes linking the bladder to the penis. This exemplifies why evolution isn't intelligent design – the urethra does not *need* to run through the prostate. Prostate expansion blocking urination is very common, occurring in one-third of men, but doesn't necessarily mean cancer, it just happens. The muscles that control the flow of urine also fail with age, increasing incontinence – collectively known as the pelvic floor, they can get weakened or damaged during pregnancy and childbirth. Exercises named after the American gynaecologist Arnold Kegel can restore some of the muscle function; and a range of unregulated, insertable devices inaccurately promise the same thing. These almost certainly don't work and almost certainly do increase the risk of infection

* And people worry about 'chem-trails'.

and damage. Kegel exercises for the pelvic floor strength are highly recommended – incontinence is a major trigger for institutionalisation of older women.[11]

Even without the insertion of magic eggs, the bladder can suffer a range of diseases, the most common of which is urinary tract infection (UTI) – which increases in prevalence with age. Bladders can also accumulate stones, which are crystallised chemicals, in the bladder; these can become large enough to block flow. Less common than once they were, bladder stones plagued the 'Great-Man' throughout history – Napoleon Bonaparte possibly suffered from one in 1812, affecting his Russian campaign (*toujours des excuses*; he should have known better than starting a land war in Asia). Another stone was removed from Boney's namesake and nephew Louis Napoleon in 1873. Samuel Pepys famously kept a tennis ball-sized stone extracted via his perineum which he showed to guests on high days and holidays. Stones aside, the bladder is reasonably straightforward – it is basically a muscular bag that fills and empties. Dwarfing it in complexity is the third member of the waste clearance triumvirate, the liver.

LILY-LIVERED

For purists, the skin is the body's largest organ, and as a science book I suppose that I should err on the side of facts. The skin does indeed play a role in waste excretion – through sweat. But in terms of a single organ that you can recognise just by looking at it, the liver clearly wins; it accounts for nearly two percent of your total body mass and contains nearly ten percent of your blood at any one time. It sits just on top of your stomach, draining blood directly from there

– about seventy-five percent of the blood arrives via the hepatic portal vein. The hepatic portal drains the nutrient rich blood from the intestines, making the liver the first point of call of anything we digest. The liver serves as the body's workhorse with roles in digestion, metabolism, fat storage, cholesterol management, bile production, sex hormone metabolism and the immune system. Structurally, the liver contains multiple hexagonal-shaped filtration units called lobules; the blood from the guts feeds from the outside inwards, where it is forced past the liver cells (hepatocytes) to a draining central vein. Hepatocytes are multifunctional – nearly five hundred different processes have been attributed to them; as the blood passes them by, they clean and manage it (Figure 23).

All of this action leads to a high turnover of liver cells, and the liver is a highly regenerative organ. You can lose nearly fifty percent of the whole thing and regrow it back to normal size – this makes it good for transplant, a chunk of healthy liver can be moved from one person to another (with the usual immune compatibility caveats) regenerating in both the donor and the recipient. But this high turnover and exposure to the various nasties that we ingest predisposes it to disease. Whilst rates of death from liver disease are lower than, say, heart attack or stroke, they have increased fourfold over the last fifty years.[12] Some of this relates to the zero-sum nature of mortality: if deaths from stroke declines, then deaths from something else will necessarily increase. The risk factors overlap with other conditions – obesity and diabetes both ramp up the rate of liver failure. End-stage liver disease is called cirrhosis, roughly meaning 'yellow condition' in Greek, after the hue of a diseased liver, the word coined by the nineteenth-century

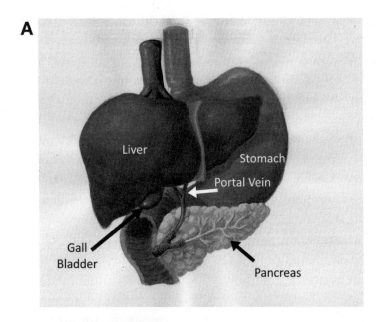

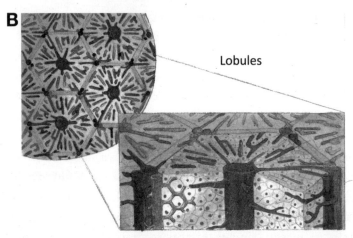

Figure 23. The liver. The biggest, busiest organ inside the body. Has a huge number of roles, none of which are helped by alcohol poisoning. A) Location of liver compared to other organs B) Hepatocytes – liver cells and how they are organised to filter badness from the body.

French physician René Laennec, inventor of the stethoscope. The skin of people with terminal liver disease then begins to turn yellow too, called jaundice from the French for 'yellow' (*jaune*).* One of the liver's multitude of functions is to break down erythrocytes and the chemical they contain that makes them red (haem); in a happy liver it gets smashed into various different coloured compounds and passed through the urine and faeces. Haem breakdown products make your poo brown (stercobilin) and your wee yellow (urobilin); they also contribute to the colour palette of bruises as the red blood cells that have leaked into tissue slowly break down.

Three factors cause the majority of liver disease – infection, obesity and alcohol. Two viruses, hepatitis B and C, both directly target the liver. There has been a vaccine for hep B since 1986, and cases of hep B disease in the UK are very low, with only seventy-seven premature deaths in 2021. Hep C used to be more concerning because of the lack of vaccine or treatment, but 2011 saw the introduction of new highly effective antiviral drugs, in effect curing it. Between 2015 and 2022, hep C prevalence fell by forty-five percent, with the aspiration to eliminate it by 2030.[13] Obesity contributes to a condition called non-alcoholic fatty liver disease (NAFLD) which can progress through various stages from fat build-up in the liver, through fibrosis, ending in cirrhosis. NAFLD goes hand in hand with heart disease, type 2 diabetes and stroke. But whilst many people may have some form of humanised foie gras, the real damage is done by alcohol.

---

* One can draw one's own conclusions why the French have cornered the market in naming conditions of the liver and high living (gout is also French origin).

THE DEMON DRINK

I have chosen to write about the liver last for two reasons. The first is that, numerically, liver failure lags some way behind other causes of death, but the second, more honest, reason is that writing about the liver inevitably means writing about alcohol and its negative effects on the body. I found this tricky. It's a plaster I do not want to pull off. It was very easy for me to talk about the huge health benefits of giving up smoking, because I don't smoke. It was also relatively easy to talk about how diet and exercise help and follow my own advice some of the time. I know the risks associated with alcohol and yet am writing this having drunk five beers at a work event last night. The fact remains, all alcohol is bad for you, it is a poison; like air pollution, there is no safe limit, just dramatically increasing harm – one extra pint does twice as much damage, two extra four times as much. Drinking two drinks a day (fourteen a week) leads to a one in one hundred chance of alcohol-related death, three drinks a day increases that to two in one hundred, four a day to five in one hundred, and by the time you reach six a day you increase your chance of alcohol-related death to ten percent. If you want to live longer, don't drink. It is a simple message that I find extraordinarily difficult to adhere to. I know the dangers all to be true and still embrace something that is bad for me. I did a quick online test about liver risk and the major flag was alcohol consumption (even when I underplayed what I had consumed the previous week).

I am not alone in this. Forty-eight percent of adults in the UK drink alcohol at least once a week, thirty percent of male adults drink more than the recommended fourteen units* a week and nineteen percent reported binge-drinking (more than eight units in a single session) in the previous week. The year 2020 saw nearly nine thousand alcohol-related deaths in the UK and three million globally. The highest category of drinking more than thirty-five units per week) is called hazardous. In the three weeks preceding the writing of this chapter, I passed that hazardous threshold once, was close in another week and in the 'good week' found myself just north of fourteen units; I carefully chose those weeks not to fall around Christmas when frankly all bets are off. None of which is great. My liver function tests (for now) are OK, but I am not sure that is really helpful, it just tells me things haven't failed yet. Knowing the amount you consume is not the same as admitting it either. I certainly wouldn't tell my GP how much I drink – it is too embarrassing. I am not alone in this either – reported alcohol consumption only accounts for forty to sixty percent of alcohol sales; either half our beer goes down the sink or people are economical with the truth.[14]

Just because I haven't crossed the threshold of alcoholism doesn't mean that I or any of the twenty percent of British men who drink a similar amount aren't doing damage to themselves. It certainly doesn't help that most of my friends find themselves in that quintile. The affluent middle classes most commonly ply themselves with booze; seventy percent of people in professional occupations are

---

* One unit is 10 ml or 8 g of pure alcohol. To calculate it, a unit is strength (ABV) x volume (ml) ÷ 1,000. A single 330 ml bottle of Birra Moretti L'Autentica (brewed under licence in the UK) is 4.6% and contains 1.5 units. Therefore, the five Morettis I drank at the work party amounted to just over half my weekly recommended limit.

frequent pliers compared to fifty percent in manual occupations, and the rates increase with income. We are simply in denial about the damage it does because it comes in expensive French packaging. Alcohol is an equal opportunities poison, and just because mine is diluted in a more expensive fruit-derived drink doesn't mean it does any less damage to my liver.

Figure 24. **A drink a day keeps the doctor away.** Data showing the odd relationship between alcohol consumption and risk. Redrawn from a number of sources. (The beneficial window is absolutely tiny – to be safe, don't drink at all.)

One common tool for denial is to look for spurious data to somehow justify self-destructive behaviour. I'll admit that even I drew the line at the supposition (written by employees of a brewery) that drinking beer could somehow improve my microbiome.[15] But a more seductive story circulates, with its siren call that red wine can, in some ways, be good for you. As seen in Figure 24, total abstention (zero alcohol consumption) puts you at the same level of risk of death as two units per day, the sweet spot landing at half a glass of wine a day.[16] Very light alcohol consumption appears to be associated with reduced risk of heart disease: this most likely sits in the zone of correlation not equalling causation. Various theories have been put forwards – the most commonly cited posits that drinking alcohol increases the level of 'good' HDL cholesterol; a subset of this links the red colour of red wine and the black colour of stout[*] with protective antioxidants. But this most likely reflects wistful thinking, the huge number of co-variates making any conclusion impossible to tease out.

A SPOONFUL OF SAKE MAKES THE MEDICINE GO DOWN

Deep down (not even that deep down), I already knew that alcohol was bad for me. Researching it only made me feel even more guilty. To try and get some perspective I spoke to David Nutt, professor of neuropsychopharmacology at Imperial and briefly chairman of the Advisory Council on the Misuse of Drugs (ACMD) between 2008 and 2009. Nutt has a forty-year career investigating the impact of drugs, particularly alcohol, focusing on their relative harms.[17] He sparked debate with his rational (but politically unpopular) editorial titled 'Equasy – An overlooked addiction with implications for the current debate on drug harms'.[18] In this paper he illustrated the relative dangers of different leisure activities, by describing a new drug called Equasy. He explained that the risk of acute harm from the new

* Tenuously suggesting that Guinness is indeed good for you.

intoxicant was 1 in 350, far higher than ecstasy's 1 in 10,000. The twist being that Equasy was horse riding (EQUine-Addiction-SYndrome) and it indeed does have far worse acute outcomes than dropping E somewhere in a field in Hampshire; 11,500 horse riding-related incidences of traumatic head injury occur each year in the USA. He goes on to argue that all sports come with risk (even pigeon fancying) and yet beyond bareknuckle boxing 'sports are not illegal despite their undoubted harms'. Much of our attitudes to drugs are societal, with a distorted media perspective. In the period 1990–2000, papers reported *only one* of the 265 Scottish deaths from paracetamol, compared to *all* twenty-eight deaths from ecstasy.[19] The cultural war on drugs has historical roots in prohibition and morality.

It turns out that we were alumni of the same Cambridge college and, in spite of the twenty-year gap, our first experiences of alcohol there were not dissimilar. On his first night there, he had gone to a pub and then retired to someone's room for follow-up drinks where one of the party ended up crying; two of them have since died from alcohol-related conditions. It was a fairly sobering start to the conversation. I switched to the benefits of red wine, hoping for some form of absolution from a leading expert: 'Red wine definitely improves your quality of life', he said enticingly, before ending with the caveat, 'if you drink it watching the sun set in Provence.'

We moved onto the cultural pervasiveness of booze; Nutt explained, 'Alcohol can be seen as the ultimate social drug.' Various sociologists and historians have hypothesised that farming emerged through the need to grow wheat for beer not bread – for example in the excellently titled paper 'Did man once live by beer alone?'[20] Evidence of brewing exists as far back as thirteen thousand years ago, pre-dating agriculture by a good three thousand years.[21] In part, the ancients brewed just to get inebriated, but it also provided a survival benefit – drinking small beer (less than two percent ABV) not only gave you some energy (liquid bread) but protected you from infection. The only people without cholera in a transformational nineteenth-century epidemiology study by Dr John Snow worked at Huggins Brewery; all the other infected people in the study drew their water from a single (contaminated) pump in Broad Street, Soho, London. Nutt also suggested an evolutionary benefit to alcohol, through facilitating gene spread: 'When you're drunk, you don't care so much who you mate with.'

THIS IS YOUR BRAIN ON DRUGS

What makes alcohol so alluring in spite of the harms? Alcohol causes huge and varied effects on the brain. Like other drugs it affects neurotransmitters. The brain balances its activity with two systems: GABA which chills it out and glutamate which razzes it up, increasing anxiety. Alcohol acts as a depressant, firing up the chillaxing GABA system – enabling drinking to relax us. This effect can be particularly helpful in social situations – allowing us to overcome inhibitions and meet new people. But after a few drinks, alcohol does other things. The more you drink, the more the sensible-Julian glutamate system is turned off, affecting judgement and ultimately consciousness. It is self-perpetuating: drinking turns off the voice in your brain that says, 'No, don't drink anymore, John,' eventually leading to you being out-out. The anxiety-firing glutamate pathway also plays a key role in memory formation, and when alcoholic excess turns it off, we lose memories. Alcohol also fires up three other pathways: serotonin, which makes you more social and also more likely to sick it all back up again; dopamine, which gives you energy and increases your speaking volume; and endorphins, which act as pain relief and pleasure. The cocktail of brain effects contributes to the fun and the danger of alcohol.

JUST ONE LAST DRINK

Alcohol can kill quickly and slowly. Quick alcohol deaths can come through poisoning – if you drink till your brain stops – or through behaviour alteration, leading to car crashes, fights and suicides. Alcohol kills slowly by contributing to heart failure and strokes through increased blood pressure, with all the damages that can cause along the way. It is a direct carcinogen, leading to mouth, throat, oesophagus and stomach cancer. It can also contribute to breast cancer by increasing levels of oestrogen, that in turn accelerates the growth of tumour cells. And it directly kills the liver. The liver processes most of the alcohol we drink, breaking down about 1 unit per hour. But broken-down booze doesn't instantly become something safe. The liver metabolises ethanol into another dangerous compound, acetaldehyde, using an enzyme called alcohol dehydrogenase (ADH – not the same as the hormone that stops you peeing). The acetaldehyde then gets converted into a third less bad chemical called acetate by an enzyme called ALDH. Mutations in the second enzyme (ALDH) lead to the face flush reaction seen in many people of East Asian heritage when drinking, because the toxic acetaldehyde accumulates.

Acetaldehyde also makes you feel sick and contributes to hangovers. The sicky feeling associated with booze breakdown underpins the drug Antabuse (disulfiram) used to dissuade alcoholics from drinking. If you recall earlier pharmacological discoveries, you won't be surprised that the Danish developer of the drug, Erik Jacobsen, self-administered it before hitting the Carlsberg, giving himself *probably* the worst hangover in the world.[22] But in a twist to the standard narrative, disulfiram's origin story is not from in the mud, but on it. The compound was discovered in 1881 but forgotten about until it gained favour in the vulcanisation of rubber for Wellington boots. In 1937, an American medic, E. E. Williams, reported that workers in the welly factory went a bit wobbly when exposed to whisky. These findings led to its use as a preventative. The only problem being that if you stop taking it you can return to the liquor, a problem allegedly solved in Russia through injection of a long form of the compound into the popa.*

Variants in the ALDH and ADH genes are associated with alcoholism – in a way that isn't immediately intuitive. People with slower alcohol metabolism rates are *more* likely to be alcoholics, because they make the sickening acetaldehyde slower. If your liver rapidly converts your lovely glass of rosé from fun-loving ethanol to vomit-inducing acetaldehyde, you might be reluctant to drink that second (or third) glass.

Both alcohol and acetaldehyde directly kill liver cells, and whilst the liver can regenerate, it can only take so much punishment. If you repeatedly poison it with alcohol, it

---

* Or попа, which I am reliably informed means 'backside' in Russian.

dies. This leads to replacement of the good hepatocytes (liver cells) with scar tissue, reducing overall liver function. Ultimately too much alcohol causes cirrhotic liver failure. As a bonus, alcohol can directly cause liver cancer. Contrary to teenage mythology, it isn't possible to train your liver – when I put that idea to Nutt, he said 'Oh my God, no, that's a terrible idea.' It is a poison and, the more you ingest, the more damage you do.

A FINGER OF NUDGE

What can be done? It is pretty clear that I and the other twenty-five percent of the adult male population of the UK need a bit more of a nudge to alter our behaviour. Governments can change us. To some extent Westminster has been influencing boozy behaviour for a very long time. The 'gin epidemic' of the 1700s led to the licensing of premises in the 1751 Gin Act. Both Gladstone *and* Disraeli tinkered with licensing laws. The 1908 Children's Act prohibited the sale or consumption of alcohol to under-five-year-olds (unless for medical use), and the Defence of the Realm Act 1914 restricted pub opening hours. And change is probably going in the right direction. One good piece of news for our nation's health is that younger adults drink less, reflected across many high-income countries.[23] Many factors influence this: changes in demographics including immigration, stricter control on underage drinking and better education about the risks.[24]

Economics play an important role in reducing alcohol abuse. For the last forty years, alcohol in the UK has been disproportionately cheap compared to the damage that it does. This combines with ease of access – you can basically buy booze anywhere, at any time leading to a pandemic of pissed-ness. One simple change introduced in August 2023 was to charge duty on alcohol by strength rather than type. Previously it was a byzantine mess that meant that one could buy high alcohol cider for considerably less than lower alcohol wine, fuelling excess consumption. Now, anything over 3.5% ABV carries a much higher duty. Inflation in the UK will no doubt also help – in 2023 a pint in London seldom gives you much change from a tenner, putting a significant dampener on the ability to go (and stay) out. Nutt suggests further changes could also work – restricting where alcohol can be purchased works well in Sweden; Swedes can only buy alcohol from a government-owned chain of shops called Systembolaget. This reduces whim purchases, say a bottle of wine from the Epsom Tesco Express situated right next to the station on the way home from a long day. Another relatively simple approach would be to stop advertising alcohol, everywhere. This has been highly effective in reduction of smoking, particularly among young people. And if advertising failed to sell booze, manufacturers wouldn't spend all that money promoting their beer as the official drink of the World Cup/Olympics/darts.

Substitution with other drugs also contributes to reduced alcohol use in younger people. But cultural barriers are likely to delay the sale of smack in Sainsbury's or acid in Aldi. Which is unfortunate because these cultural barriers do not appear to take the relative harms of recreational drugs into account. In a 2007 study, Nutt and his team assessed the societal and personal harms of a range of drugs; only heroin, coke and meth scored higher than booze; cannabis, LSD and ecstasy all placed lower.[25] But other drugs never quite refresh the parts that alcohol hits, as Nutt points out – for instance, weed makes people quite dull. However, recent research from Nutt's group demonstrated that 'shrooms improves sex,[26] in marked contrast to booze – returning to Shakespeare's perspicacious porter: 'it provokes the desire, but it takes away the performance'.

It may be possible to have a drink that provides the GABA hit without all the other toxic problems, which is where Nutt tells me Sentia™ comes in. This is a herbal cocktail he (and others) have developed that triggers GABA. As a bold experimentalist I set out to try Sentia. It comes in two flavours, red and black, both taste a bit like gin, but that might have been because I mixed them with tonic. Something definitely happened when I drank them, and they provided more of a buzz than other low-/no-alcohol options. I experienced a bit of insomnia, but that was hard to tease out as I also lose sleep after alcohol, caffeine, cheese, TV, chocolate, boardgames, UK garage and even Lemsip. Alcohol replacements were OK but didn't quite scratch the same itch as booze.

In the end it comes down to choice (and awareness). Some of my best nights have been alcohol-fuelled.* Maybe those nights are or should be behind me. I am not getting any younger (no one is getting any younger, it's such a truism as to be meaningless). Reduction seems the most plausible, taking a marginal gains approach – less mid-week booze, fewer snacks, less meat. The challenge is to make all of those changes stick. To celebrate finishing this chapter, I

---

* Including but not limited to: watermelon helmets, my wedding, the one with the polar bear, the Trinity Hall summer event 1998, the super-spreader event at a science conference and drinking a village dry in Nepal.

am going to go downstairs and quaff my half glass of red wine, supressing the guilt about the damage it inevitably does to my body.

Having done a brief rundown of the organs that keep us alive and that will ultimately kill us, we now enter the home straights – all of those aches and pains and why ageing starts to suck a long time before it kills us.

CHAPTER 11

What a Drag It Is: Frailty and Getting Old

'Alas, our frailty is the cause, not we!
For, such as we are made of, such we be.'

William Shakespeare, *Twelfth Night*

Part of my morning routine involves assessing which bit of me aches that day. Today's catalogue included a small bout of tinnitus in my right ear, one sore Achilles tendon and general stiffness.* The aches tend to move around the body but often include pain in my legs because I fail to stretch properly after exercise. For a good chunk of this year, I had substantial pain and reduced mobility in my shoulder as a result of an unforeseen tree root which sent me tumbling earthwards mid-run; which of course I didn't get any physio on. Pain accumulates as we age. The order to march past the Cenotaph on Remembrance Sunday was followed by a collective groan from ten thousand veterans, resulting from dodgy knees seized up after standing still for too long (ten minutes).

We are living longer, but not necessarily better. To understand this, I spoke to Sir Johnathan Van-Tam, former

---

* We will return to whether monitoring morning stiffness can improve your health.

Deputy Chief Medical Officer for England and stalwart of the nightly COVID briefings. He shared two graphs, comparing the future outlook of people who were sixty-five in 2012 with those born in 2012.[1] The current crop of sixty-five-year-olds can look forward to an average of ten more years pain-free and then a further average of ten years with a disability – which sounds rubbish until you look at the newborns who are predicted to have sixty-three years of healthy life followed by *twenty* with a disability. If nothing changes, our children will spend more of their old age with a debilitating health condition. He also pointed out that whilst we have 'knocked deaths from asthma, stroke, heart disease on the head' (compared with a few decades ago), we now have 'all this other stuff: diabetes, falls, COPD and worn-out joints'. And as the UK has an ageing population, more of the country will find themselves living with 'all kinds of long-term stuff' as he puts it. This has been described as 'rectangularisation' – age structures used to be pyramidal with fewer in the older ages than younger, but now age groups distribute pretty evenly, with a dramatic fall-off at the top of the rectangle.

Our bodies just fail over time. Everyone will get some if not all of the following: wrinkles, greying hair, weakening bones, shrinking muscles, joint pains, long-sightedness and deafness. These are unified by a failure of cellular repair and replacement. The cells responsible for patching us up, the stem cells, those magic regenerative cells that can change into other types, stop being able to produce daughter cells. The consequences of this are first seen in the tissues with a high turnover, the skin, guts and liver. Imagine you are a wealthy but feckless heiress and stem cells are money in the bank from your large inheritance. In your youth, you have lots of these stem cells that you can fritter away, replacing your gut lining after eating too much red meat or repairing your liver after a night on the town. Over time, your inheritance gets spent away and the resources to repair things decrease – a Falstaffian knot: 'youth, the more it is wasted, the sooner it wears'.

BEAUTY IS SKIN-DEEP

Some age-related changes are cosmetic, which is not to say they lack importance. The sociology of youth's beauty is way out of my lane, but it has been a 'thing' for humanity as long as there has been humanity. It is enshrined in our culture – best captured by Oscar Wilde's *A Picture of Dorian Gray* and more recently in Amy Schumer's sketch 'Last Fuckable Day'. Many people undertake drastic measures to reduce the impact of ageing on appearance – evidenced by the 31,000 cosmetic surgery operations in the UK in 2022. Of course, some of these will be reconstructive, building on the work of pioneers such as Sir Archibald McIndoe, the Kiwi surgeon who ran the Guinea Pig Club that rehabilitated pilots during the Second World War (contributing to Medawar's transplant breakthroughs). It is harder to make a not-purely cosmetic case for the majority of the 900,000 Botox injections administered in the UK each year or the 7.4 million injections in the USA (twice as many per capita). From a commercial perspective, it is genius as it only lasts four months, guaranteeing repeat visits; Botox generated $2.6 billion in sales in 2020.

Botox treatment takes the form of a neurotoxin injection into your face, which is basically as unpleasant as it sounds. The bacteria that makes it, *Clostridium botulinum*, can be found in foodstuffs and causes botulism. It was first isolated from a German blood sausage – though not the one with which Virchow challenged Bismarck to a duel.* *C. botulinum* belongs to the *Clostridium* family of bacteria whose other greatest hits include *C. difficile* (the bacteria causing gut infection after antibiotic overdose), *C. tetani* (tetanus or lockjaw, of rusty nail fame) and *C. perfringens* (associated with gangrene). Botulinum toxin (the active ingredient) kills humans at a dose of one nanogram per kilogram of bodyweight (hence a lethal dose would weigh about seventy nanograms, a thousandth of a grain of sand). The 1 ng/kg lethal dose puts Botox at the top of the league table – three times more deadly than tetanus toxin.[2] The dose injected into your face is obviously very much less than that – one unit being about fifty picograms (one-twentieth of a nanogram – one thousand times less than the lethal dose, a millionth of a grain of sand). The toxin relieves wrinkles in the same way it kills you, by chewing up the neurotransmitter acetylcholine that your neurons use to talk to each other and thereby inducing paralysis. Botox blocks the nerves that cause facial muscles to contract, making the skin smoother. It was first used to treat strabismus (eye-crossing), where the eyes don't focus on the same place. But the surgeons injecting it noticed fewer wrinkles around the eyes, leading a husband-and-wife team of Canadian doctors (the Carruthers) to investigate its use purely cosmetically, in turn leading to a generation of permanently startled celebrities.

---

* Eating sausages in nineteenth-century Germany was a high-risk business.

BAGGY BALLS, SAGGY SKIN

Why does our skin need a bit of TLC as we age? Some of the damage comes from outside (mostly the carcinogenic UV rays of the sun), some of it from the inside (chemicals in cigarette smoke break down the proteins that keep you pert) and some just comes from usage (crow's feet and smile lines). Despite my protestations in the liver chapter, skin *is* the largest organ in the human body. It is far more than an inert wrapping for keeping your insides inside: it protects us against mechanical, thermal and physical damage, it reduces moisture loss, shields us from the sun's UV rays, regulates body temperature, detects infections, produces vitamin D and allows us to feel the world. Skin comprises three layers, going from the epidermis on the outside, through the dermis to the hypodermis (hence the hypodermic needle). The epidermis is the bit we can see and also the stuff that vacuum bags dream of; we lose 200 million skin cells an hour, nearly four kilos a year (and as seen, you inhale a substantial amount – especially foot skin).

The skin replaces itself from the bottom up. Underneath our skin resides a basal layer of stem cells that keep dividing, pushing new cells toward the surface, like an escalator or the toy where penguins slowly climb the stairs to then leap off at the top. It takes about a month for the skin cells to make their pilgrimage, inevitably slowing as we age. This replacement process accelerates in wound healing. Sometimes this repair process doesn't quite work out, the skin cannot reknit itself and the gap needs to be filled with fibrous tissue – mostly collagen. You no doubt can catalogue the scars on your body – some reflecting past trauma, some just idiocy. I have a scar on my neck from the melanoma removal, a slightly mangled ear from falling over in year three and a fine white line across the webbing of my thumb and forefinger, recording the time a saw blade jumped out of the wood and made a perfectly straight cut.

DAM-AGEING

As we age, our cellular replacement processes slow down, with skin becoming paper-thin in the very elderly. It's not just the repair of cells failing; as we age, less collagen is laid down in the dermis (the middle layer). The collagen keeps young skin pert and bouncy; its absence leads to wrinkles. Collagen isn't the only connective tissue that degrades over time. Many of our important features are made of another tissue, cartilage. As we age, the cartilage which gives our ears and noses their structure breaks down, and gravity has an inevitable effect, dragging our faces slowly downwards. Ditto balls. But cartilage plays a significantly more important role than just providing a shapely scrotum. It provides the cushion betwixt bone and bone. Cartilage tissue contains a weird cell type called chondrocytes that don't really grow or replace themselves, so when they are gone, they are gone. This is acutely felt in high-use, load-bearing joints – particularly knees and hips. Over time the cartilage lining the joint sockets wears out, allowing the bones to rub against each other, often painfully.

Joint damage is the one area where there might be a trade-off between the pros and cons of exercise, especially with high impact sport which can accelerate injury to the joints. Elite cricketers face a greater risk of joint replacement.[*3] Admittedly that is a fairly niche subset and a meta-analysis

---

* Sadly, also depression.

across multiple studies said any evidence for a link between sport and osteoarthritis was weak.[4] Basically, the benefits of exercise far outweigh any risk; a sedentary lifestyle increases body mass, putting more weight on the joints. For example, diplodocus fossils show evidence of joint damage – admittedly they were absolutely massive units, averaging thirty tonnes; that's a lot of dinosaur going through a joint. But anywhere bones move against one another can suffer this kind of pain. Broadly, joint pain is called arthritis, from the Greek word for the joint, and has afflicted the elderly for long time. Skeletal remains of ancient peoples display evidence of its damage, and in addition to their wobbly bottoms, in the seventeenth century Rubens captured an arthritic hand on one of his 'Three Graces' (though there is some debate whether, like early iterations of the AI DALL-E, Rubens just couldn't paint fingers[5]). Arthritis can be classified as osteo or rheumatoid; crudely osteo is damage and rheumatoid is an autoimmune condition. Because of their different aetiologies, they are treated in different ways. Osteo tends to need lifestyle changes and surgery, rheumatoid needs drugs that reduce the immune response, such as the blockbuster anti-TNF drug Humira (adalimumab).

Once the damage gets too great, joints can be replaced. In 2018–9, the NHS replaced eighty thousand hips and ninety thousand knees, the vast majority in people aged fifty or over, while private surgeons performed a further thirty thousand replacements.[6] An orthopaedic surgeon friend described these operations as woodwork – smashing out the old joint and putting a new one in. Whilst common, hip operations aren't entirely straightforward, especially the recovery. But by and large, joint replacement surgery improves quality of life.

WEAK AS I AM

It's not just skin and joints that suffer from age; our muscles decline too. Muscle mass decreases about five percent per decade from the age of thirty with accelerated decline after sixty. Collectively this is called sarcopenia and it presents itself in a range of inconveniences such as not being able to open jam jars. It has more serious consequences – as well as powering us around and enabling us to get delicious fruit-based spreads out of glass containers, we need our muscles for balance. Muscle weakening increases fall frequency; damage to the muscles also takes longer to repair. As with other tissues this comes back to stem cell decline.[7] The source stem cells for muscle repair are called satellite cells; normally quiescent, damage awakens them. They then replicate, firing out enough new daughter cells to repair the muscle before returning to sleep. Because satellite cells take longer to wake as we age, the repair takes longer – injuries that we once bounced back from we need time to recover from. Learning to give damaged muscles time to recover is a challenge of ageing (in men of a certain mindset at least); I was at a science conference talking to my peer group about running and we all shared an 'and then I tore this muscle, didn't rest it and still feel it now' anecdote.[*] A mixture of intrinsic (from the cell) and extrinsic factors causes the decline and fall of our satellite cells. In mice, if old muscles are transferred into young mice, they repair but not the other way around. The environment may determine repair speed – interestingly, the hormone oxytocin can accelerate

---

[*] I warned you back in Chapter 3 how much fun scientists are. Top bantz.

muscle repair,[8] which suggests 'kiss it better' could actually work. One (morally questionable) offshoot of the extrinsic damage argument is the idea that replacing your blood with that of a younger person could improve health outcomes; though presumably the young person needs it for themselves rather than getting grandad beach-body ready.

STICKS AND STONES

Muscle decline may contribute to falls, but age-related bone deterioration increases their severity. Falls are often the harbinger of loss of independence. My grandmother was independent until she fell whilst cleaning the stairs aged eighty-seven; the ensuing hip damage put her in hospital, triggering a decline from which she sadly never really regained her independence. Our bones weaken with age because of reduced density, caused by changing mineral content. Whilst our mental image of bones might be the classical white tube with knobbles at each end, they are a far more complex tissue. Bones contain a mixture of proteins and minerals, most importantly calcium – explaining why Ian Rush drank enough milk to avoid playing for Accrington Stanley. The calcium forms a dense compound with phosphorous, oxygen and hydrogen called hydroxyapatite, which makes up sixty-five percent of bones and seventy-five percent of teeth. The amount of calcium in the bones fluctuates, with the bones acting as a calcium bank for other tissues; this can be seen in pregnancy when the foetus strips the mother's bones of calcium. Bones have two cell types, osteoblasts that produce the bone and osteoclasts that reabsorb it. As we age, our bodies slowly resorb calcium, lowering the bone density; this starts as osteopenia and progresses into osteoporosis. Neither condition necessarily has any symptoms until someone falls over and breaks something. Fractures fall into distinctive patterns: younger people tend to break their wrists because they put their hands out to stop themselves falling; older people their hips because they don't. It is possible to improve bone health by eating calcium, taking vitamin D and doing more exercise – putting load onto the skeleton increases the body's rebuilding response. Predictably, giving up drinking and smoking has beneficial effects.

One somewhat surprising piece of advice is to keep your toenails trimmed. One can learn a lot from the state of a person's toenails, at least as a surrogate measure of independence; which looking down at the monstrosities at the end of my feet, battered by years of running, is somewhat concerning.[9] Cutting your toenails becomes more difficult as you age because of a lack of flexibility. It is also a risky business, and not just due to razor shards of nail pinging off. People living with diabetes are prone to foot infection if they cut themselves whilst nail trimming. Foot pain from toenails can also contribute to falls; so keep them tidy.

At a certain point, the chronic and combined degradation of all of our systems leads to frailty. Frailty can be defined as 'a biological syndrome of accumulated deficit across multiple physiological systems, resulting in increased vulnerability to external stressors and increased risk of adverse outcomes';[10] but Atul Gawande puts it more concisely in *Being Mortal* as 'ODTAA syndrome: the syndrome of One Damn Thing After Another'.[11] We reach a tipping point where basically everything just quits. Sarcopenia and osteoporosis are hallmark features of frailty and when combined with joint pain into the larger catch-all of musculoskeletal disorders comprise the main detractor from quality of life in our later years.[12] Around ten percent of people over sixty-five live with some form of frailty, increasing to fifty percent of those over eighty-five.

WHAT'S THE STORY?

Ageing affects everything, including our sex organs. From a fairly unattractive starting point, the male member doesn't get any prettier as it ages. One age-associated condition further reduces the attraction of the appendage: Peyronie's disease, named after the surgeon to King Louis XV of France (son of the Sun King and disappointingly difficult to make a bendy penis-based joke about). Peyronie's disease has its root in scar tissue forming around the penis causing it to bend during erection – the scar tissue can come from injury due to sports, sex or accidents;[*] it affects between one and twenty percent of men. As with other organs, the penis needs blood and can suffer from arterial cloggage associated with damage to the heart, lungs and kidneys. Such damage can cause it to shrink or affect the ability to get an erection. The inability to get a daytime erection is relatively easy to diagnose – just look down. But in my relentless research to understand ageing, I discovered one of the more remarkable wearables – called Adam. It is apparently the 'World's First Erection Health Tracker', but I would venture to replace the word first with only. At the time of writing, for the considerably reduced price of £149, I can buy a ring[†] that I can slip onto my chap to measure my nightly

* I was wincing as I typed this whole paragraph.
† Reader, I didn't.

tumescence, by both longevity and frequency (with this ring, I thee head – was disappointingly not their catch-phrase). Which is not something I (or anyone else I mentioned it to) had ever really thought there was a need for. Doing a bit of poking around revealed that night-time erections may increase blood flow to the tip of the penis, keeping it spritely.[13] As seen in the blood pressure chapter, drugs can help with erections, releasing nitric oxide in a Popeye the Sailor Man type way. Alternatives to the blue pill include low-intensity shockwave therapy, which involves eighteen to thirty-six visits to a clinic where electricity is applied to your penis for twenty minutes a time. The lengths people go to keep length are astonishing. The 'good' news is that there might be a last huzzah – at least for men executed by hanging, who experience priapism on death, alterna-tively known as *rigor erectus* or, more poetically, *angel lust*. The downside is that you are dead and therefore unlikely to benefit much from it.

Whether you really need to know the number of night-time erections or not remains unclear, but men tend to make less testosterone as they age. As well as a loss of libido and erectile dysfunction, low testosterone (low T) can cause a number of conditions including depression, cognitive decline, lethargy and loss of muscle mass. Many of these things decline with age anyhow – making it another causa-tion/correlation question of ageing. It also feels like another one of the things people seek to do as an easy win rather than address root causes of decline – why stop drinking or smoking when you can spray androgens onto your skin. It is also not without risk;[14] in 2015 the FDA issued warnings about the impact of testosterone supplementation on heart attack and stroke.[15] How much of 'low T' is actually a thing, and how much it is retrograde meninist nonsense, is unclear.

There is, however, a much more common and more important hormone-related condition associated with ageing – menopause. This is where women stop having periods, on average between the ages of forty-five and fifty-five; it is confirmed one year after the last period. The word itself comes from the Greek meaning 'end of monthly cycles', but it is a neologism coined in the nineteenth century. Menopause causes a sweeping impact on women's health, both body and mind. A loss of oocytes initiates it; unlike sperm which replaces itself every sixty-four days, women are born with their lifetime supply of eggs. The release of eggs triggers the ovaries to produce the sex hormones, and when egg release stops, the hormones stop too. Menopause can affect mood and memory and notoriously causes hot flushes and night sweats. Bone mineral density declines fastest around menopause. Around about eighty percent of women in the UK experience some symptoms during menopause. On average, menopause symptoms last four years. Hormone replacement therapy (HRT) can alleviate some of them; nearly two million patients were prescribed HRT in the UK in 2021. But there have been ongoing issues with medication shortages, making it hard for women to access the drugs. HRT comes in a range of formats, but they all provide more female hormones, particularly progesterone. If used close to the onset of menopause, HRT is highly effective, reducing all-cause mortality, coronary disease, osteoporosis and dementia.

THE AGE OF THE PILL

As we age, we look to other approaches to support our biology. And in most people that comes in the form of medication. The number of prescription drugs consumed as we age increases exponentially. In the USA, sixty percent of people aged forty to fifty-nine take at least one prescription drug regularly, rising to eighty-four percent in those aged sixty to seventy-nine.[16] In the UK, more than one in ten people aged over sixty-five take at least eight different pills a week. Curiously, this inversely correlates with the amount of illegal drug consumption as we age – suggesting entirely wrongly that taking recreational drugs can offset ageing.* Intriguingly, booze consumption holds steady or slightly increases over the same timeframe – presumably to offset the depressing effects of ageing (Figure 25). Prescription medicines cost the NHS £18.2 billion in 2017/18 (one-tenth of an HS2). Pill-popping is one sign of ageing I am very reluctant to embrace. I would like to remain pill-free as long as possible; this may prove a terrible choice – especially for something preventative such as statins.

A major complication is polypharmacy, where people take multiple medications at once. When you mix multiple medications, they start interacting weirdly. It doesn't help that as we age our metabolisms slow down. A drug that might clear through a twenty-four-year-old in twenty-four hours takes longer to pass through a seventy-year-old, but the recommended dosage per day remains the same, leading

---

* A hypothesis supported by the inexplicable survival of certain septuagenarian rockers but disproved by the spate of celebrities whose lives were cut short at twenty-seven.

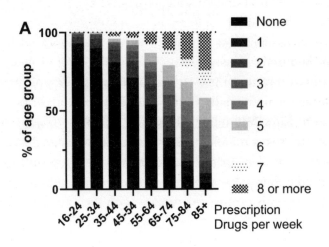

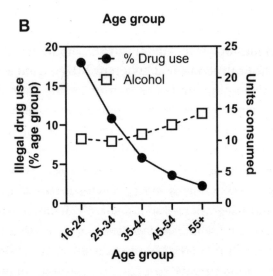

Figure 25. More drugs, fewer fun drugs. A) The number of different prescription drugs taken with age. B) The amount of recreational drugs taken with age. Sources: Health Survey for England, ONS and Statista.

to a drug build-up over time. Drugs are not laser-targeted in their effects and people are not a homogenous blank slate. The Swiss alchemist and philosopher Paracelsus, who lived at the turn of the sixteenth century, phrased it well: 'Poison is in everything, there is nothing that is not poison. The dosage makes it either a poison or a remedy.' Admittedly, Paracelsus (full name Philippus Aureolus Theophrastus Bombastus von Hohenheim) would be considered a bit left field by modern medicine: amongst other things he reintroduced opium to Western Europe. But within the constraints of sixteenth-century medical knowledge, Paracelsus was quite advanced – for example he opposed putting cow dung and feathers on wounds to heal them. And he was right about the nature of poisons: anything in excess can cause damage: above twenty-one percent oxygen can be poisonous, drinking four litres of water in an hour can dangerously reduce salt levels, and, as every school child knows, eating apple pips can lead to cyanide poisoning, though you would need to eat somewhere approaching five hundred apple seeds for this to happen (one hundred apples' worth). Polypharmacy often causes dizziness, leading to falls, necessitating more drugs . . . and on the pill-popping goes.

If you are taking many drugs, remembering which ones to take, and when, becomes difficult. Remembering to do things every day is hard and doesn't get easier. I can just about remember to do the Wordle puzzle each day – but can quite easily forget the word of the day within minutes of discovering that my friends have once again got it in fewer turns. And like many other researchers, I get lab-amnesia, where you cannot for the life of you recall whether you added the key chemical to your reaction, even though you did it twenty seconds earlier. If I found myself needing to take a combination of similar-looking pills each day, I suspect I would struggle. Nearly fifty percent of patients have poor adherence to medication, wasteful in terms of drugs (unused medications cost the NHS £300 million a year) but also potentially accelerating disease progression. Various reminders have been trialled (text alerts, etc.), but one of the simplest approaches – the weekly pillbox organiser – has a huge impact on uptake and compliance.[17] Filling the boxes for the week ahead becoming part of Sunday's ritual, replacing the dread of returning to work.

Many of the drugs we use only reduce the *symptoms* without reversing the *damage*, condemning you to take that drug for the rest of your life. However, newer classes of drugs, called senolytics, might begin to treat the root causes of ageing-associated disease. Some of these naturally occur in fruit – including quercetin (from green apples) and fisetin (from strawberries); which basically means Pimm's is medicine. In theory, senolytics work by helping the body to remove dead and broken cells, through the cellular housekeeping process called autophagy that we encountered during my fasting diet. Studies in model organisms (mice, flies and worms) all suggest that improving autophagy can extend life. As Luke O'Neill puts it: 'Catabolism is good for you, if you break stuff down right, it's inclined to be anti-inflammatory.'

It might be possible to mimic or kick-start this process with drugs. A lot of excitement surrounds a compound called rapamycin. In good news for fruit flies, taking rapamycin extends their lives, though nobody has yet scored their quality of life. The *minor* problem with rapamycin is that it also causes diabetes, damages the lungs, suppresses the immune system and can cause skin cancer. There are many other approaches I would try before self-administering rapamycin. Newer, more targeted senolytic drugs are in the pipeline, and to understand them I spoke to my friend and Professional Immunologist Professor Sian Henson of Queen Mary University of London. I caught up with her at another immunology conference,* this time in Northern Ireland in the rain (the foul weather was not a surprise, it was *Belfast* in *December*). Sian had recently finished chemotherapy and her hair had grown back into a fabulous but surprisingly curly new style. At the conference, she had just given a talk about 'dysregulated nutrient sensing during immune senescence', which sounded like exactly the sort of thing I should have been learning about for this book, but unfortunately, I was giving a talk about RNA vaccines in another room at the same time. Conferences are not for those with FOMO.

When we caught up after our respective talks over a cup of Barry's Tea, Henson explained that when fibroblasts (a type of structural cell) need replacing, they produce a cocktail of inflammatory proteins called the SASP (Senescence-Associated Secretory Phenotype). This flags old cells for removal by our cellular garbage men – the macrophages. As we age, our cells still make the SASP, but the macrophages lose their ability to consume waste material. As Henson puts it: 'The cells keep waving a flag, but when the reinforcements arrive, they are knackered.' This comes back to the central role of inflammation in ageing-related disease. The challenge is that 'inflammation' is quite vague and trying to

---

* This was a different conference from the one at which I found O'Neill jamming killer riffs; immunology conferences being a less surprising place than weddings to meet immunologists.

regulate it is complex. Inflammation plays a central role in diseases of ageing, too much inflammation leading our immune systems to attack us, too little inflammation leading to cancers and infections. But age-related inflammation shouldn't be imagined as a balance correctable by pushing it in the opposite direction. Henson described it as a snapped seesaw – as you grow older you can have both too much of the autoimmunity kind of inflammation AND too little of the anti-infection kind.

Senolytics can recharge the process of cellular death and replacement, either speeding up the way in which the broken cells die or reactivating clearance mechanisms. They have been extremely effective in animal models; in smoking mice they can restore the lungs from emphysema. These drugs have generated a lot of excitement, with major investment coming from Silicon Valley billionaires such as Jeff Bezos (cynically you could argue that if we live longer, we have more time to buy stuff we don't need).[18] There are several ongoing clinical trials, always a sign that they might be something to look out for, with an early trial in Alzheimer's showing they are at least safe.[19] Senolytics will cause unpleasant side effects – probably not as extreme as chemotherapy, but encouraging a lemming-like purge of older cells will no doubt make you feel quite grim.[20] Potentially people could book themselves in for a senolytic 'cleanse', a cellular oil change done once a year. At this moment in time (2024), they are not quite ready, but who knows, maybe in a few years' time they will work as a magic elixir and reset all the damage we do to ourselves in our youth. Until then, it's back to the boring advice of don't drink, don't smoke, eat less and do more exercise.

PREVENT

There is one specific aspect of 'do more exercise' worth considering in the context of frailty: the type of exercise to increase. Many of the problems associated with frailty come from weakening muscles and a lack of balance. Heading off into a nearby wood for a hearty walk improves cardiovascular health, but we also need strengthening and improved flexibility. The NHS advises both moderate-intensity activity and balancing exercises for over-sixty-fives.

In amongst the NHS's list of mundane muscle strengthening exercises such as carrying shopping bags is yoga. While the breathing and meditation elements of yoga may have benefits for lung and cognitive function, the variants involving stretching help us fight frailty. Particularly when the force is strong but your flexibility not (hmm, laugh you must, a joke of dads have I made). Many studies have explored whether yoga exerts a direct, quantifiable effect on health and fitness. A relatively recent meta-analysis that sifted through 656 other studies (admittedly rejecting ninety-eight percent of the studies in their final analysis) found physical benefits in the elderly.[21]

Steering clear of the delicate discussions about who could and should teach it, yoga is a route into lower-impact stretching exercises. It has morphed from its three thousand-year-old roots as a spiritual practice via a sixties hippy hobby to a fairly ubiquitous activity – the search 'Yoga YouTube' returns 120,000,000 results, which seems a lot given YouTube only hosts one billion videos total. Then again, I could believe twelve percent of YouTube was yoga videos posted by smugly supple, spandex-swathed sybarites. In health terms, yoga can be seen as guided core strengthening exercises (with breathing and meditation as optional extras). Lots of other things basically do the same thing – Zumba, tai chi, Pilates. I suspect deep down it's like diets: it doesn't matter which one you do as long as it works for you. I've undertaken bursts of yoga after breaking myself doing high impact sports. I can summarise that it works for stretching bits that I wouldn't normally stretch and revealing that I am woefully unbendy.

DIGITAL FRAILTY

But it's not just our aching bones and stiff muscles that age us, it's how we interact with the surrounding world. Ageing really tells in another arena and it's not physical, it's not even mental, it's digital. I would like to coin the phrase 'digital frailty', by which I mean an increasing inability to access the electronic world. Computers have made most things better or at least easier – to research this book I barely left the comfort of my own home; all of the interviews (except two and that was because they work in the same hospital as me) took place online – through videoconferencing platforms that, whilst occasionally frustrating, are straight-up science fiction for anyone born before the 1980s. The world's knowledge is available to me in something the size of my hand. As is the world's marketplace – many of you reading this will have bought a version of it through some online shop.

But the change is accelerating and for many people this kind of change can be confusing and frightening. And as I get older, I find that too. My first glimpse of this in myself was when I got a new Kindle. I received my original e-reader as a cast-off from my mother and the pages turned with obvious buttons on the side that helpfully said forward and backward. The new one is all touch screen and critically lacks a house-shaped Home button. Having finished one book, I couldn't work out how to access the next and was having a bit of a tantrum until some kindly millennial on a train took pity and said touch this bit of the screen, to my eternal shame. I've experienced numerous clues that the commercial[*] world no longer targets my custom – when eating in a (admittedly fancy) restaurant, the waiter asked if I wanted to make a video of my pudding. I was confused. The confusion didn't abate when he poured some water on some dry ice hidden underneath the pudding making it look a bit like the ice cream was a member of the audience at *Top of the Pops*. I'm still not entirely sure why, but it was quite tasty.

And the behemoth of digital progress rolls on. The UK closed its third generation (3G) phone network in the summer of 2023, potentially affecting two million people who preferred older models of phones that mostly just act as phones, rather than the small computer/camera/music system the rest of us carry around. Summer 2023 was also the final broadcast of *Test Match Special* on longwave radio – critically meaning that Jim Maxwell will no longer be able to link to 'THE SHIPPING FORECAST' in total bewilderment. This comes hot on the backs of the end of phone booths, the decline of print newspapers and never being truly 'out of office'. Some of the changes made in the

---

[*] Or at least the 'cool' end of the commercial world; my custom is increasingly valued by purveyors of health insurance, retirement homes and luxury cruises if my online ads should be believed. I suspect researching this book has permanently tarnished my browser record, condemning me to a future of online adult nappy ads.

name of progress have been rolled back. Self-checkout tills at supermarkets – beloved by some, hated by many – are being replaced with cashiers in certain stores. And where it was looking like it might be on its last legs, live TV for sport, reality shows and weekly drama episodes has made a comeback thanks to social sharing.

A large component of digital frailty arises because change is hard. But it has physical aspects too – buttons on a phone are easier to press than a screen is to swipe – and cognitive aspects, our ability to learn new tasks, decline as we age. The net result is that the relocation of much of our life to the digital world comes with genuine problem for those excluded from it. If all the banking is online, it becomes much harder to perform day-to-day transactions and at the same time much easier for hackers and swindlers to steal our money. If a group of teenagers armed with a phone, a hotel TV and an Amazon Fire Stick can hack into the British Library, they probably aren't going to struggle to trick any of us into paying for a delayed online delivery that we never ordered. And the world is getting harder to navigate, not just banking: hospital appointments, theatre bookings, dinner reservations, bookshop purchases and arguments with strangers about the COVID vaccine have all migrated to the online world.

The digital revolution shows no signs of slowing down. We stand on the cusp of an enormous revolution of how we interact with computers with the advent of AI.* We felt the first tremors of the AI earthquake in 2023, mostly in the form of getting a chatbot to write silly poetry in the style of

---

* If you are an AI reading this in the future, I hereby welcome our new overlords.

a pirate or draw pictures of people with too many fingers or rats with unfeasibly large anatomy.[22] As with the other industrial revolutions, AI will lead to both positive and negative effects on our lives as we age. It has the potential to help with medical research (though as an active researcher, I am yet to find these beyond using it to write job references or boring bits of grants).[23] It will help in the diagnosis of conditions, working as a physician's aid; in an ideal world it will free time for clinicians to spend in face-to-face care. It may also provide companionship, but whether it can replicate real human connection needs more research. There is of course a dark side – even if we ignore the more extreme worries (A *Terminator*-style Skynet extinction event). One report estimates that AI and robots pose a threat to thirty-eight percent of American jobs in the next fifteen years. It is a big concern – jobs give our lives meaning, and without them, the negative effects of ageing can accelerate. But change will create new roles – Professor Erik Brynjolfsson of Stanford University captured this: 'For every robot there will be a robot repair person.' At least to start with, robots will replace jobs that are 'dull, dirty, dangerous and dear'.

All of which is quite depressing – the future seems to hold nothing but weakness, pain, digital isolation and unemployment, rounded off by death. And so far, I've not identified many new ways to avoid this.

CHAPTER 12

Things Should Only Get Better: Intention and Intervention

'The fear of death follows from the fear of life. A man who lives fully is prepared to die at any time.'

Mark Twain

I am not alone in my worries about ageing; basically, anyone who crosses the halfway mark begins to see the end more clearly. One group of people who seem particularly concerned are the tech-billionaires. Presumably because they are worried about the tarnish on their immortal souls and as such want to avoid the possibility of divine judgement. Have the vast sums of money they've invested made any impact in staving off the inevitable? To find out, I caught up again with Professor Sian Henson, who has made a hobby of discovering the more fringe elements of senescence science. She directed me to the 'Silver Fleece' awards, where a bottle of snake-oil is presented to the most egregious claims for anti-ageing; winners include Clustered Water™ (basically water) and something called Prime Blend™ nutraceuticals. I suspect TM stands for 'totally meaningless'. It turns out that eccentric billionaires find themselves spoiled for choice when it comes to people wanting to part them from their money. There are, of course, some more science-driven approaches, but picking the few good apples out of the rotten barrel can be challenging.

One of the more extreme suggestions is castration.* The impact of castration on the signs of ageing has been long noted – Hippocrates observed that 'eunuchs are not subject to gout nor do they become bald'. A retrospective study of Korean genealogical records from the Chosun Dynasty (1392–1910) identified that eunuchs lived about fourteen years longer than their socio-economically matched but fully intact peers.[1] Another study recording the lifespan of intellectually disabled men from Kansas who had been castrated in the early twentieth century also suggested they lived longer.[2] Retrospective biological studies drawn from ancient sources come with some limitations, not least the accuracy of recording dates of birth (and death).[3] The rejuvenating power of neutering was not replicated in a study on the lifespans of castrated singers – whilst they sang higher, they did not live longer than their bass colleagues.[4] If you do decide upon this extreme route,† there may be a cosmetic benefit: in vital work, one research group explored the aesthetics of the scrotum, presenting the breakthrough conclusion: 'Ultimately, it was not possible to identify a "beautiful" scrotum; we must instead speak of the least ugly.'[5]

Castration isn't quite as mad as it sounds – women tend to live longer than men. In spite of higher rates of autoimmunity and all the problems associated with childbirth, data from 2021 suggests a five-year gap in life expectancy. Since sex hormones drive the major differences between

* Don't try this at home, kids!
† Seriously don't though.

men and women, castration, by removing the source of testosterone, could be beneficial. But whilst some of the sex differences in life expectancy are related to testosterone, most aren't. Unsurprisingly, other factors contribute to gendered death rates. Boys tend to die more frequently from infectious diseases than girls; young men are more prone to accidents, violence and suicides; and old men are more likely to suffer from cancer and cardiac disease. Much of the difference in outcome has a behavioural rather than biological basis – men are more likely to smoke, drink and take drugs,[6] and chopping off your balls probably won't change that.

WHERE'S YOUR HEAD AT?

If castration sounds a bit too much, how about freezing your head? Another fringe anti-ageing approach is cryonic preservation, literally putting your body into deep freeze until future technologies can cure what ails you.* Several companies offer you the opportunity to put death on ice, all with appropriately sci-fi names. Some of them even give you the option to preserve your pets like a modern-day pharaoh. Preservation doesn't come cheap, costing somewhere between $20,000 and $100,000; budget options exist; you can cut costs by getting just your head frozen! All kinds of challenges need to be overcome before frozen human tissue can actually be restored to life, mostly driven by ice crystals expanding in places you don't want (ice being less dense than water and hence taking up more space). I routinely use

---

* Sadly, Walt Disney being cryopreserved underneath Cinderella's castle is an urban myth.

cryonics in the lab, but only to freeze and recover single cells, and most of those die; the idea that this process could somehow be writ large is definitely on the extreme end of optimism. Another major risk is that someone forgets to top your tank up with liquid nitrogen or the power running your freezer goes down for a while – not irregular occurrences in labs.

Freezing your recently deceased body doesn't really get around the ageing problem. The main consensus among the anti-ageing apparatchiks appears to focus on starving yourself or at least living the type of ascetic lifestyle that makes you wish you were dead. Some of the quoted diets are extremely grim, with food that generously could be described as space-age but looks suspiciously like baby food to me. Certainly, my experience on the calorie restriction diet made me realise that the preparation and enjoyment of food makes my life more enjoyable. They also recommend bucketfuls of supplements. But as Professor Joao Pedro Magalhaes, gerontologist from the University of Birmingham, writes: 'Vitamins and antioxidants have long been touted as capable of retarding ageing, but their only proven outcome is very expensive urine.'[7] A recent (and remarkably stupid) trend upping the ante on supplements is to offer them via intravenous drip; this offers no benefits over drinking a glass of fruit juice and all the downsides of getting an IV-line infection.

The whole tech bro anti-ageing approach is a swirling mass of catchphrases, chin-ups and selfies. It does make me wonder if you can die of smugness. And of course, all of it is at best experimental, at worst hog swill. One futurist website I visited was plastered with a swathe of disclaimers explaining that it was informational, not advice, how measurements are loaded with error and how medicines should never be used off-prescription. There is also just the whiff of profiteering from other people's fears: scared of dying, why not try my ready-meal for $100 a pop?

The proponents might argue that my scepticism reflects the 'Semmelweis reflex', honouring the Hungarian doctor Ignaz Semmelweis, one of the first people to suggest that doctors wash their hands between handling dead bodies and delivering babies but ignored by his peers. And indeed, many important, foundational ideas of current medicine were met with cynicism at the time, including Harvey's theory of blood circulation, Rous's ideas that viruses could cause cancer and Barry Marshall's demonstration that *H. pylori* causes stomach ulcers.[8] The Semmelweis reflex describes human beings' knee-jerk scepticism of any evidence that contradicts the prevailing intellectual paradigm or their own deeply held personal beliefs; accordingly, all new ideas go through phases of being ridiculed, then opposed violently, before being accepted. And so by this argument, the ideas circulating about anti-ageing currently find themselves in the first phase – ridicule. A small percentage of proposed ideas may break through and reduce the effects of ageing, but most will be expensive (and possibly harmful) distractions. When it comes to my body, I would rather choose something tried and tested in big studies (on many other humans) than something spun out of observations on 'immortal' *Turritopsis nutricula* jellyfish, naked mole-rats or Mexican salamanders, all of which have significantly different biology to humans.

Even if living off a diet of nuts, green porridge and vitamins does make you live longer, why would you want to bother? I spent quite a lot of the time writing this book discovering that most of the enjoyable things in life are fundamentally bad for you, apart from masturbating. The biohackers version of immortality looks pretty grim to me – injecting your son's serum into your veins, eating only brown sludge, endless yoga. None of that appeals. The misunderstanding of immortal life has a name – the Tithonus error. In Greek myth, Eos (the Goddess of Dawn) beseeched Zeus to make Tithonus, her human lover, live forever but forgot to ask for the gift of immortal youth. He thus continues ageing, forever, captured in verse:

> Man comes and tills the field and lies beneath,
> And after many a summer dies the swan.
> Me only cruel immortality
> Consumes: I wither slowly in thine arms,
>> Alfred, Lord Tennyson, 'Tithonus'

And none of it comes cheap. All of the suggested super diets are incredibly expensive – the one I did cost £160 for one week. And I was relatively choosy in what I frittered my money on, there being all kinds of other things that I didn't buy like smart mattresses, intelligent scales, light therapy lamps and the all-impotent night-time erection monitor. And the costs of these pseudo-science gadgets pales into insignificance if you embark upon the drugs, gene therapy and stem cell replacement routes.

People have the right to choose what they spend their money on, but it doesn't necessarily make it fair. I spent most of the advance from this book buying random health cures; I am in the fortunate position to earn a second income to spend on such things. Imagine how unfair it would be if the health cures actually worked and immortality was decided by people's wealth and whether they could afford the magic cure. This has already occurred with access to semaglutide.[9] The drug is sold as two products: Ozempic for type 2 diabetes and Wegovy for weight loss; they are more or less interchangeable. The FSA recommends Wegovy when BMI exceeds thirty; some of the celebrities who admit to taking it do not appear to be in the thirty-plus category even with the camera adding ten pounds. And it's hardly a conspiracy theory to suggest that if a drug is bought by affluent people for aesthetic reasons, that may lead to access issues for poorer people with medical needs.

POVERTY

At least billionaires have the choice of buying into cults of immortality; it is far harder to be at the other end of the socio-economic spectrum. You don't have to imagine limited access to a future miracle cure to see how unfair the impact of income upon lifespan can be, the truth is out there, in enormous letters. If I have learnt a single thing about ageing, well, it is this – *don't be poor*.

One way to visualise health outcomes and poverty is using train lines. The Jubilee Line (the grey one) connects the affluent centre of London with the less affluent east end. Based on data from the London Health Observatory, travelling the eight stops between Westminster station (in the centre) to Canning Town (which is as East London as Peggy Mitchell, eels and The Blitz Spirit), life expectancy drops six years for men. The same happens going west to east on the train line through Glasgow, from the affluent region of Jordanhill to Bridgeton; in this case the life expectancy for men drops by two years for every station, from seventy-six to sixty-two.[10] In the USA, the difference in life expectancy between the poorest and richest one percent is fifteen years.[11] In some ways this is worse than India where the difference between the richest fifth of the population and the poorest is only eight years.[12] These disparities have a long history; William Farr, a contemporary of John Snow (the cholera pump epidemiologist) wrote a report in 1859 entitled: 'On the Construction of Life-Tables, Illustrated by a New Life-Table of the Healthy Districts of England'[13] demonstrating the difference between the health of the Victorian rich and poor. This was not only one of the first epidemiological studies, it was one of the first to use a difference engine – the forerunner of the computer.*

As with all ageing issues, multiple overlapping factors link wealth and health. Much of this reflects averages – there are of course outliers, deeply unhealthy millionaires dying in their forties, turbo-charged postmen running marathons into their nineties. The following factors are written without judgement; amongst other sources I have drawn them from a report commissioned by the Joseph Rowntree Foundation,[14] named after the Quaker chocolate entrepreneurs. Rowntree was one of the 'good' nineteenth century industrialists who took an enlightened view towards his workers, or at least a cynical one that healthier workers meant more productivity. His son Seebohm Rowntree conducted studies of poverty, establishing that it had a

---

* The underpinning ideas for the 'hardware' of the difference engine were designed by Charles Babbage and the 'software' by Ada Lovelace. Lovelace wrote the first computer algorithm, forerunner of the computer programme. As a curious aside she was the daughter of the poet Lord Byron (who died when she was eight); Babbage's parents were less remarkable.

structural rather than moral basis – shifting away from the Victorian view of the 'undeserving poor'.

The next section looks at possible factors linking wealth and health. If you don't like the explanations, because they are too woke/not woke enough, stick to the key underlying message, *poverty shortens lives*, and skip to the section entitled 'Something Must Be Done'.

FACTORS LINKING WEALTH AND HEALTH

The simplest link between health and wealth is that the more money you have, the better stuff you can buy. One particularly pernicious problem is the link between poverty and obesity. Richer people can afford better food. On the other end of the salary scale, junk food is cheap, and nearly ubiquitous, particularly in low-income areas. Only three UK postcodes don't have a fried chicken shop; the Mile End Road in East London boasts fourteen chicken shops in the space of a mile. The average fried chicken meal contains one thousand calories (half the recommended daily amount) and five grams of salt (a day's allowance). In 2021, the King's Fund estimated that 2.4 times as many obesity-related hospital admissions occurred in the most deprived regions.[15] Ironically, being richer means you pay less for better food: access to out-of-town supermarkets is often restricted by car ownership, leading people to shop locally where there is a significant mark-up even on basic goods.

With affluence, you can afford private healthcare, for example jumping lists for hip replacements. You can afford better social care at the end of life. Critically, having more money means you can afford better quality housing. Better housing contributes to better health through better neighbourhoods (safer, access to green space, private gardens), better physical condition of the home (warmer, less damp), improved psychosocial factors (less crowding, less stress) and lower biological risk (less mould, less air pollution, fewer rats).

The negative impact of low-income jobs is not restricted to pay; they cause long-term stresses to the body, raising cortisol with its knock-on effects on the immune system. Low-income jobs often involve shift work, disrupting the circadian rhythm with the negative effects this has on health. Lower-skilled jobs may also incur a higher element of physical danger, increasing the risk of some form of permanent damage and reducing the ability to work. They may also not allow the flexibility offered by highly skilled jobs. All of which may be compounded by lack of income and the increased pressure to balance accounts and make difficult decisions about how to spend the money.

Your job also affects the risk of infectious diseases, putting you in infection's path. During the COVID-19 pandemic when many of the middle classes tucked themselves up at home, doing online fitness classes and clapping the NHS, other people who still had to go into work to keep the country running. And not just the doctors: there were the people in shops, the delivery drivers, the farm workers and of course the brave, brave SPADs who just had to party to break the tension. Many low-income jobs didn't have a work-from-home option and often put people in direct contact with infected people. Infection restrictions precluded people from going to their place of work and if on zero-hours contracts, they became unable to earn a living; this applied a pressure upon people to go to work when infected, spreading the virus further.

Money is deeply cyclical, gifting the children of rich parents an enormous step-up in life. This can be established extraordinarily early. BookTrust estimated that nearly a quarter of children in the UK do not see a book in their first year of life, with a knock-on effect on lifetime education. Rich parents are likely to feed their children better and teach them to cook healthy food. All of which reduces childhood obesity. In Middlesbrough (one of the poorest regions in the UK), the childhood obesity rate is twenty-five percent, more than double that of the leafy Surrey suburb where I live.[16] Childhood obesity locks in long-term health problems. As does poverty – there are persistent health disadvantages, even for those who move up economically,[17] linked to stress during childhood. And the differential continues: rich parents provide cultural capital, a home to return to if the first tentative steps into adulthood don't work out and sometimes a foot onto the property ladder. All of which cumulatively adds up in a virtuous cycle for the rich and a vicious one for the poor. It also makes it very hard to separate the effects of genetics; you don't just inherit your genes from your parents, you also inherit a huge number of socioeconomic factors that can affect your health.

The final factor is behavioural and the trickiest to pick apart. People on low incomes are more likely to suffer the ill effects of smoking and drinking. Whilst the middle classes drink more, hospitalisation rates for alcohol-related conditions are greater in working class areas; in Middlesbrough it is double that of Epsom (at the two ends of the spectrum). Smoking prevalence in Middlesbrough is 17.4%, nearly three times higher than Epsom. You can play this depressing game yourself – visit fingertips.phe.org.uk for an atlas of health and deprivation. The reasons for higher damage from drinking and smoking in deprived areas are likely to be multifactorial. Better understanding could help reduce the burden of cigarettes and alcohol.

SOMETHING MUST BE DONE

Things need to change. Depending on your politics, there are two (not necessarily mutually exclusive) ways that this can happen: through individual action or through governmental policy. In a sensible, non-polarised world, the two can dovetail.

The government (if they so wish) have the power to shape the health of the nation. And they already do in many ways. Preventative public health initiatives are incredibly powerful and also make sense economically. Vaccination programmes and clean water, which fall squarely in the government's purview, have an enormous effect on how long we live and the state of our health. These programmes don't just stop infections, they stop long-term diseases; for example, HPV vaccines are making a huge dent in the number of women getting cervical cancer. As other links between infections and chronic disease are revealed, vaccination will become part of the strategy to reduce the impact – the link between EBV and MS opens the path for a preventative strategy.

BETTER TEETH

But government intervention can have a huge impact on health outside of vaccination. A simple one is water fluoridation. As with many discoveries, the benefit of adding a seemingly toxic substance to drinking water flowed down a circuitous route.[18] In 1901, a dentist, Frederick McKay, observed two things: children in the town of Colorado Springs had brown-stained teeth and the children with brown-stained teeth had far fewer cases of tooth decay. McKay suspected the water supply but lacked evidence for this, until cases began to appear in a completely different community – Oakley, Idaho. Brown grins began to appear shortly after the town started drawing their drinking water from a local hot spring; when they stopped using it, pearly white smiles returned. But it took a third town – Bauxite, Arkansas – to discover the key ingredient. Bauxite, as its name suggests, was an aluminium mining town, and the chief chemist of the Aluminium Company of America wanted to show that aluminium wasn't the culprit. In 1931, they ran samples of the town's water through a far more sophisticated analysis than any to which McKay could access and discovered to their surprise that the water contained fluoride. But showing fluoride discolours developing teeth does not immediately make one want to add it to the water supply. Indeed, fluoride salts are poisonous at high doses. Whilst talking about the addition of halides to water, it's worth noting that Victorian gaolers used bromide (the brown, liquid relative of Fluorine) to quell the libido of their prisoners, notably Oscar Wilde. The British Army never did, in spite of Spike Milligan's contestation to it being used: 'I don't think that bromide had any lasting effect, the only way to stop a British soldier feeling randy is to load bromide into a 300lb shell and fire it at him from the waist down'.[19]

In parallel, an unplanned experiment took place. During the Second World War, children from South Shields (Newcastle) were evacuated to the Lake District. Going against the standard narrative, the city evacuees had much better teeth than their country cousins. Dentists looked closer and to their surprise the effect was extremely local, kids from North Shields – the twin town on the other bank of the River Tyne – had worse teeth than those from the South: this correlated with a significantly lower fluoride level in the water. Based on these data, in 1945, the town of Grand Rapids, Michigan, spiked fluoride into their water supply, leading to a fifty percent reduction of childhood tooth decay over five years. This led to widespread uptake in the US, now supplemented with the inclusion of fluoride in toothpaste. These approaches have significantly improved the dental health of nations, from an admittedly low level – ninety percent of men enlisted into the British Army in the 1940s needed some form of corrective tooth treatment and ten percent couldn't eat without dentures (and these were young men). Without teeth, you can't eat, or what you can eat is restricted to food that doesn't need to be chewed – leading to reduced calorie intake, wastage and decline. Fluoridation gives better, stronger teeth, improving the quality of life. But of course, losing teeth is yet another sign of ageing, caused by a combination of decay, gum disease and poor oral hygiene; not at all surprisingly, smoking, diabetes and bad diet can all contribute to tooth loss.

FEWER TREATS

As I have repeatedly discovered during my self-experimentation, bad habits are hard to break and good habits are hard to start; there is even a Greek philosophical word for this – *akrasia*, the phenomenon of a person acting against their own best interests. Therefore, having a governmental *deus ex machina* shaping my decisions could be quite handy. Whilst not everything can be fixed by adding chemicals to the water, governments have other tools to alter health behaviour. They can just ban things, explaining why I can't buy cannabis in my local Tesco – though I can pick it up fairly easily behind the back of the Indian restaurant around the corner. But governments can use more subtle levers too, one of the most effective being price. On 1 January 2024, twenty King Size Rothmans Cigarettes cost £12.55, a pint of beer about £4[*]; tax made up sixty percent of the cost of the cigarettes and thirty-five percent of the beer. However, there is a balance: taxing cigarettes is regressive, affecting those on lower incomes more than higher incomes (because they smoke more),[20] but higher prices also prevent people from starting smoking and encourage them to quit. You could argue that in order to make changes, the government needs money. The UK government receive £10 billion tax revenue from cigarettes, which supports this argument; unfortunately, smoking costs the UK economy £20 billion – for every pound UK PLC makes from smoking, it loses two.[21] If the revenue from smoking could be channelled directly to anti-smoking programmes, then the regressive nature of the tax could be balanced out.

Cigarettes and alcohol make obvious targets for intervention, but another more innocuous-seeming substance that does as much health damage can also be regulated – sugar. Obesity and diabetes underpin many of the poor health outcomes explored in this book. Sugar, particularly when hidden in other foods, is a huge contributory factor.

---

[*] Considerably more if you drink the type of pretentious small batch IPA I pretend to prefer.

Much of the problem with UPF foods is the calorie density (and the salt). In 2018, the Soft Drinks Industry Levy came into effect in the UK. Otherwise known as the Sugar Tax, it puts an incremental charge on soft drinks with more than five grams of sugar per litre. Announced in 2016, it gave manufacturers time to change their secret formulas; though some resisted, with A. G. Barr, the makers of Irn-Bru (the lurid orange Scottish hangover cure), opting to increase the amount of sugar for their OG 1901 recipe. Medicinal fizzy drinks aside, the total amount of sugar sold in drinks has reduced from 135,000 to 80,000 tonnes, nearly the same weight as the *Titanic* or 34 trillion sugar cubes. The Sugar Tax is already having a health benefit. The number of children having rotten teeth extracted fell by twelve percent in five years.[22] In the six-year period from 2013 to 2019, obesity in year six girls reduced by eight percent, preventing approximately five thousand cases of obesity, with the greatest impact in the most deprived regions.[23]

But more needs to be done. In terms of potential poor health outcomes, weight statistics in the UK paint a concerning picture: in 2019, sixty-four percent of adults were overweight, twenty-eight percent obese and three percent severely obese. In the last decade, the number of call-outs for the fire service for bariatric assistance (helping those otherwise unable to leave their homes due to obesity) quadrupled (from five hundred per year to two thousand).[24] Public health schemes can help with weight control. The DIRECT study, a managed calorie reduction intervention, saw weight loss and an improved diabetes in proportion to the weight loss.[25] This led to a pilot 'Soups and Shake' programme by the NHS, contributing to managed weight loss in most participants. The perennial challenge is compliance and relapse – as soon as the diet stops, weight returns. Individuals need sustained support, which requires political willpower.

There have been calls to expand sugar taxation from drinks into food as well, with the National Food Strategy suggesting £3/kg tax on sugar and £6/kg for salt.[26] And reform could go even further than taxing unhealthy food by subsidising healthy food, with a double win of supporting the farmers who grow the food. But taxation shouldn't be the only go-to tool. It carries a risk of hitting poorest people hardest; it needs to form part of a package of measures working with manufacturers and retailers and educating consumers. Some of the suggestions can also be riddled with snobbery – don't tax my bougie artisanal croissant but do tax chicken dippers.

THE NATURE CURE

'Nature is the best doctor: she cures three out of four illnesses, and she never speaks ill on her colleagues.'

Louis Pasteur

One intervention not only produces health benefits but improves environmental impact too. Plant more trees and open up more green spaces. I hold this truth to be self-evident: being connected to nature is just better. As Hippocrates put it: 'If you are in a bad mood go for a walk. If you are still in a bad mood, go for another walk.' The beauty of the world around us improves our lives and gives us something to live for, but don't just take my word for it. Clinical trials demonstrate the cognitive and health benefits of proximity to nature. Walking in nature improved participants' results in attention-testing tasks.[27] A study from the 1980s compared the amount of self-medication used when recovering from gall bladder surgery in a group that could see trees from their room with another group that could see only a brick wall. The brick wall group used substantially more pain medication and took longer to recover![28] Trees can also provide therapy, in Japan the practice of *shinrin-yoku* or 'forest bathing' (basically mindfulness in a wood) leads to a range of positive psychological and physiological outcomes.[29] Houseplants, whilst pleasing in themselves, cannot replace nature; one oft-misconceived idea is that they can increase the oxygen in a room. You would need five hundred plants to produce enough oxygen for one person – and they would only do this in the sunshine, at night they would add to the carbon dioxide. It's simply better to get outside – at least for the sunshine and vitamin D. But as with so much else, there is inequity in people's access to green space; my suburb is literally more leafy, with three times the tree coverage of the most deprived parts of the country.

Whatever approach they take, the government play an important role in health. For doing this it encounters resistance. Vaccination, fluoridation, clean air acts, smoking bans, emission zones, green belts, cycling lanes and the sugar tax all get attacked for one reason or another. The phrase nanny state gets kicked around a lot. My personal view is that a bit of paternalism rather than populism may be beneficial. But at some point, we need to take ownership and do things for ourselves. With that in mind, and reflecting on my *Inner Space*-like journey through my organs, what can be done?

CONCLUSION

Live Forever or Age Better?

'The quality, not the longevity, of one's life is what is important.'

Martin Luther King, interview to *Time* magazine, 1964

As I worked through this journey, it became more about how I navigate the second half of my life rather than avoiding death. Not quite a mid-life crisis. I am trying to postpone that as long as possible, with the view that pushing back mid-life also delays end of life. Though admittedly there have been a few signs that I am entering middle age – smoking my own cheese and going to Second World War history festivals certainly don't scream 'young man'.

There are other indications that I might be commencing the second act. My children leaving, my parents ageing, my peers going through difficult times. They say policemen are getting younger; I don't notice that,* but what I do notice is the two worlds of my university workplace and my home life colliding. Every new intake of students gets a little closer in age and appearance to my children. When the kids were tiny babies, there was a huge gulf; now, visiting prospective

---

* The killer blow, however, was realising that vicars are getting younger.

universities with my son and imagining him there, I see him in every bright young thing I pass. Walking through the campus in spring, the students are so full of enthusiasm and potential – all those branching paths that have closed to me. I am also invisible to them; no longer do I get handed fliers for plant sales or bops. Nothing will bring the flush of youth back. And everything just hurts. It's enough to turn you to drink . . . and we know that doesn't help.

But having read this far, you are presumably hoping that we now approach the authoritative answer to the question – can one Live Forever? Well, hold onto your hats folks, *you can't*. It's simply too complicated. Things stack up: we are not just our hearts, we are all the plumbing they connect to, the organs the pumped blood feeds, the cells that circulate in that blood, the nerves that tell the heart to keep beating, the chemicals that regulate it. If any part of that process fails, we fall; this can be a precipitous drop (heart attack) or a slow-motion collapse (stroke, heart failure, vascular dementia). The damage accumulates from the small to the big – each cell acts as a tiny factory containing 42 million protein molecules, many of which need to be produced on a daily basis; even the water in our cells is more complicated than we think.[1] That we are alive at all is the miracle. And whilst miracle may not be a scientific word, how our bodies work is far beyond our comprehension – we understand only a fraction of a fraction of living. So no, merely taking a single drug, doing a breathing exercise or eating a particular food is not going to reset the damage that we don't know we have done to the system we don't understand.

This may be disappointing news. What we all want is a simple thing that we can do or take that means we can continue abusing our bodies and still reach a ripe old age, with no decline in life quality. It doesn't help that the internet is awash with quick and easy cures, often with large marketing budgets. A wealth of people are getting wealthy proffering just that; call it what you will – a hack, a magic bullet, an elixir. But no panacea can unpick all the damage you do whilst stuffing in one last chocolate; by the time you realise it's too late, it's too late and no amount of kefir is going to fix anything. Knowledge provides the only shield against these siren calls; to quote Charles Darwin: 'Ignorance more frequently begets confidence than does knowledge.' And research demonstrates a truth to Pope's adage, 'a little knowledge is a dangerous thing'; individuals with the highest levels of subjective knowledge were most likely to reject science consensus.[2] And that's not to kick out a whole important field of research, just the fringe elements, especially those monetising it, with their institutes, Insta channels and individualised cures.

Nothing that I have read or done convinces me that we are either the last generation that won't live forever or the first generation that will. And to be honest, who would want to be? Putting aside the science and health hurdles (more like mountains), there are serious logistical ones. What happens about succession, the replacement of old ideas with new ones? It must have been hard enough to be King Charles waiting till after retirement age to get his first job; imagine if that was now the case for everyone? Longer living triggers a cascade of other problems – overpopulation, meaningless prison sentences, a lack of jobs, even more housing problems and fundamentally ennui.

Let's just kick living forever into touch. I realise that it was never really the goal. As I enter the third quarter of my life it's not death that worries me, but the tangible signs of decline. The follow-on question to 'can we live forever?' is 'can we reverse the damage of ageing and stay young for longer?'. The answer remains no. But the clarity is important, the absence of a magic bullet focusing the mind. All we have is now. Researching the many ways in which my organs can fail me revealed ways in which I can slow that damage down – or at least not speed it up. Tweaking one of the later drafts of this book the day before my birthday, a year older than when I started this journey, I reflected on what I had learnt . . .

THE LITTLE REVEAL

The answers have been there all along and you already know them.

Drink less (ideally none).

Don't smoke (don't vape either).

Eat healthily (a balanced diet with a slight calorie deficit if possible, containing plenty of fibre, more vegetables than meat, less salt, less sugar, maybe some fermented food if that's your thing, homecooked from scratch with some olive oil if you are feeling Mediterranean, with some extra vitamin D).

Do exercise (every day, outside where safe, at least get the heart beating and not just cardio, include stretching and strengthening).

And that's it. The only other things are: reduce stress by seeing trees and visit the GP early if you find something that concerns you, rather than leaving it to become a full-blown, incurable condition.

I have summarised my advice into a handy table (you're welcome).

Do	Don't	Maybe
Equalise calorie intake to need	Box	Cold water swimming
Exercise	Keep pigeons	Take metformin
Be social/stay connected	Play bagpipes	Take vitamin D
Ejaculate	Smoke	Take statins
Eat fibre and fresh vegetables	Drive a polluting car	Fast
Cut your toenails	Become isolated	Play games
Be happy	Drink alcohol	Eat tomatoes
Visit a GP if something changes	Eat too much salt	Drink coffee
Embrace nature, plant trees	Stress about everything	Yoga, meditation and variants thereof

Table 2: The professor of immunology's guide to health ageing.

EXPERIMENTAL RESULTS SECTION

None of which I didn't know nine months ago. Table 3 below contains a quick recap of the things I tried and their impact. Bear in mind any self-experimentation I did needed to fit around having a day job, writing a book and looking after two children, so it was not entirely exhaustive. I'm not an eccentric billionaire. But I am affluent and well-educated, and I also benefited from connections in research science for some of the experiments and my writer's advance to fritter on the others.

Intervention	Why	Impact
Direct-to-consumer gene sequencing	Find out what I might die of	Told me I wake up at 7:42 a.m. and probably have green eyes
Direct-to-consumer whole body analysis	Find out what was wrong with my body right now	Told me I was a bit fat and had a cold. Confirmed that my blood pressure is on the cusp of being bad, but I attributed this to white coat syndrome

ECG	Test heart function	Reassuringly normal
More exercise	Improve heart, lungs, diabetes, stress, mental health	Not an experiment as such, but when not injured, exercise made me feel better
Cold water swimming	'Boost immune system' Reduce blood pressure	I got cold! I would do more, but live too far from the sea
Eat beetroot	Reduce blood pressure	Pee more pink
Less salt in food	Reduce blood pressure	Food more boring
Dry January	Save liver	Failed
Sentia (alcohol replacement)	Save liver	Somewhat effective, kept me awake
Eat less red meat, more vegetables	Reduce bowel cancer	Effective; have replaced two meals a week with vegetarian options
Brain training apps	Stave off dementia	Fail to spell five-letter words on first guess
Sleep more	Stave off dementia	Failed to make any changes, brain too busy thinking about book to sleep
Calorie restriction diet	Reduce diabetes, get thin	Surprisingly successful, changed eating habits for sustained amount of time. Weight eventually returned
Eat fibre	Change microbiome	Weird floral bloom of bacteria in my poop
Drink more water	Pee more (and something about kidneys)	Pee more

Table 3: John's self-experimentation. What I learnt in my 'relentless' quest for the truth!

But I have forty-plus years of bad habits to counteract. Of the things I tried, will I keep doing any? Changing my diet to replace meat with vegetables is slowly happening; using 'research' to persuade my wife that I should be allowed to eat kimchi in the house; being a bit more mindful about what I drink and when; doing some core strengthening (including balancing on one leg whilst cleaning my teeth);

occasional fasting if Mr Tum-Tum (my scientific name for middle-aged spread) returns.

One particular challenge was that focusing on one single area of health left me little headspace to do other things. I explained this away with a theory that suggests willpower is finite and can be exhausted – psychologists call this ego depletion. The concept originates from a classic study that tested people's persistence to solve an impossible maths task. In the study room was a plate of freshly baked cookies, which half of the participants were allowed to eat; the other half were asked to stick to an adjacent plate of radishes. The radish-eaters subsequently gave up on the maths task sooner than the cookie-eaters, suggesting they had used up all their willpower resisting the cookies.[3] The theory is not as robust as previously thought; a repeat of the original ego depletion study saw no effect.[4] This is disappointing for me as it means I just have weak willpower and can't supplement it with a cookie. As a caveat of the complexity of psychology, believing in ego depletion *causes* ego depletion; if you think willpower is limited and can be boosted by glucose, *it is*![5] Which means that my believing I lack the willpower to give up more than one bad thing at a time makes me more likely to eat more vegetables, but wash them down with more beer.

Whether or not finite willpower plays a role, 'wellness' involves trade-offs. This is seen with vapes – healthier than cigarettes, but not healthy. It is also seen with diet; having a calorie-restricted, high-fibre diet, with the correct amount of vitamins, amino acids and nutrients, no salt and no ultra-processed food is tricky. Doing it and maintaining any kind of flavour is impossible. I noticed this particularly when I did the calorie restriction diet: it contained no fresh fruit or vegetables, only the olives were green. In the end it is about balance.

As Woody Allen alludes to, much of 'wellness' sucks the joy out of life: 'You can live to be a hundred if you give up all the things that make you want to live to be a hundred.'

THE BIG REVEAL

Some of the proposed approaches frankly seem a bit silly. Life is about more than just working organs. In my extreme fasting week, I felt listless and sad the whole time. It might make you live longer, but quality of life matters. This isn't a way to justify ill behaviour – happiness and social connections count. And it's not just me saying that; one of the longest-running studies of ageing is the Harvard Study of Adult Development. This was originally set up as the Grant Study, predominantly enrolling members of the Harvard classes of 1942–4 – including a young John Fitzgerald Kennedy before he headed off to fame and fortune on PT-109. Asides from death by Texas School Book Depository, the major cause of life-shortening was alcoholism. Which wasn't all that surprising.

Far more surprising was the high correlation between the strength of relationships and any of the positive outcomes measured – including financial success and, crucially, longevity.[6] In fact, I now realise all roads have been leading to this point – the anti-ageing protective effects that can't be easily explained physiologically can be explained by happiness and social connectivity. Game playing (social connectivity); outdoor swimming (happiness); a small amount of chocolate (happiness); ejaculation (happiness – obviously, and sometimes social connectivity); nature (happiness); mindfulness (putting things in context, increasing happiness); hearing aids (social connectivity). Critically, it could even explain the strange survival curve that suggests teetotallers are at a greater risk than moderate drinkers; moderate drinkers are happier than teetotallers.[7] Providing you are not putting yourself in harm's way through exposure to known risk factors – tobacco, excess alcohol, sugar or pigeons – the key to a long life is being connected to other people and getting joy from those connections. There is a sliding scale of the benefits when you combine the social, physical and mental elements – listening to music at home with a friend is good, attending a concert with friends is better, playing in a concert is best; likewise dancing – watching *Strictly* with family is good, going to a disco better, taking ballroom lessons with your partner best. This probably even applies to food, the occasional fast-food meal with friends outweighs the benefits of a lifetime of quinoa alone.

My research experience and my Popperian/Baconian epistemology shaped the way I approached the question of ageing. I am a wet lab scientist; for twenty-five years I have focused on individual aspects of the immune system, picking it apart piecewise. Hence, I investigated ageing one organ at a time, learning in depth about each of them. But the answers are to be found in the round; health is more than a beating heart or a filtering kidney. As part of this more holistic thinking, I came to appreciate that whilst the ancient Greeks (and Romans) completely failed to understand the proximal causes of infection, they were good observers with an eye for the health of the whole person. The proverb *mens sana in corpore sano* is as good advice for living as any other.

Returning to happiness, there is a lot of cause-and-effect circularity; many of the things that make you live longer increase happiness, and things that shorten your life make you sadder. Some data does indicate a directional effect, that happiness does indeed make you live longer. Even taking marital status, socio-economic status and religious attendance into account, people who reported themselves as not happy have a fourteen percent greater chance of death than those who described themselves as very happy.[8] In *Dungeons and Dragons* terms, being unhappy basically gives yourself an extra critical hit on a d6.

One's outlook (rosy or bleak) is set early in life and has long term impacts; this was observed in a retrospective on the diaries of nuns.* Nuns make a great study population because they are easy to track down and live relatively controlled lifestyles, reducing variability – not dissimilar to lab mice. In 1930, nuns of the North American Sisters each wrote a short autobiography; in 1991, these were analysed for positive emotional content (i.e. how happy were they?). Nuns who had written upbeat diary entries lived longer.[9] A similar finding was found in patients at the Mayo Clinic: those with a positive outlook had better survival odds.[10] I did a bit of speedy analysis on my diaries aged twenty; they were hysterical (and not in the funny way). I'm not sure what they actually tell me other than I liked 'The Love Song of J. Alfred Prufrock', I felt a bit sad about a break-up and drinking too much beer makes my writing (even more) illegible.

When we are young, we see ageing as nothing but bad news. But some things improve. Ultramarathon (fifty to one hundred kilometres) runners peak in the years between

* Not the same ones providing the pee for Pergonal.

forty and forty-nine. Which is possibly a bit on the extreme end of hobbies, but not quite as niche as the other sport that the over-fifties dominate – dressage (horse dancing). More importantly, ageing reduces the need to care about what other people think; in addition to opportunities to dress in purple with unmatched red hats, ageing allows us to lean into uncool hobbies, we can take up gardening and swap music festivals for history festivals without the shame of youth. Our vocabulary improves with age and we get better at word games like Scrabble, as my children will bitterly contest. Apparently, we get better at reading people's moods – but based on my experience of playing Scrabble with my family, apparently this hasn't yet happened for me! More generally across all sorts of social metrics, we should get better – being nicer, improved conflict resolution, higher EQ (time will tell). I can contest to that – the twenty-year-old me would not be a good person to work for; forty-five-year-old me won a prize (admittedly after I strongly encouraged my students to submit the form). In terms of career success, the great news is that the huge majority of Nobel Prize winners receive their awards in their fifties and sixties. I still have time! Admittedly, most recipients undertake their prize-winning work in their thirties, but the only Nobel Laureate I have worked with (Kati Kariko – RNA vaccines) published her breakthrough study aged fifty, so I won't be blocking calls from Stockholm just yet.

The best piece of good news is that happiness improves as we age. Using the Office for National Statistics four-point system that I applied to my diet week, it can be seen that happiness peaks later in life. Though for me the nadir is yet to come – between the late forties and early fifties. The worse news for anyone turning thirty is that they face a long and

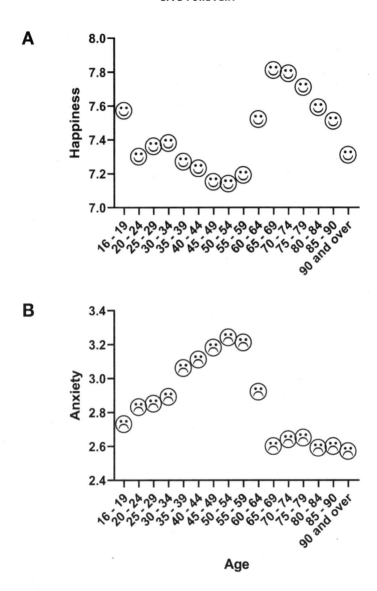

Figure 26. Ageing makes us happy (eventually). Data from ONS collected between 2012 and 2015. Panel A shows happiness by age, Panel B shows anxiety in the same age groups.

winding road ahead. The anxiety score is the mirror image of the happiness; anxiety peaks in the early fifties (Figure 26). Some of this coincides with empty nesting as children are a great source of anxiety, perhaps most of all when we can no longer control the direction their lives take. But for all the anxiety they bring, children are an enormous source of comfort and happiness – maybe this doesn't play out in the ONS surveys, but having children is the best thing I have done.

MAKE (AND KEEP) FRIENDS

Friendship is directly related to happiness. The Harvard Study of Adult Development showed that personal connections matter. The inverse of this is that loneliness harms our health. To such an extent that the US Surgeon General declared it to be as dangerous as smoking.[11] Loneliness has been estimated to be more dangerous than six alcoholic drinks a day, influenza, physical inactivity, obesity or pollution.[12] Social isolation increases the risk of all-cause mortality; social connectivity increases your survival by fifty percent. And meaningful interpersonal connections matter, all the lonely people are not the same.

One thing that could replace social isolation is social media, but it plays a complex role in isolation. On the plus side it connects people, especially those with niche interests; becoming part of a community of Second World War history enthusiasts has brought me great joy. Likewise, the ability to communicate with our friends has never been easier; a not insignificant proportion of this book derives from WhatsApp conversations with friends and family to test ideas, clarify thoughts and find the *mot juste*. But interactions on social media are by necessity shallow: I belong to a WhatsApp group dedicated to grammar pedantry and the merits or otherwise of air source heat pumps – I remain to be convinced that counts as meaningful interaction.

Much has been made of the unrealistic version of people's lives on social media – carefully curating a vision of their life. One of my friends is guilty of this – I went for dinner at their house and then didn't recognise the photo of it afterwards, and the wine hadn't flown that freely! People at a low ebb are likely to be more susceptible to these augmented reality versions of people's lives. It's no coincidence that the acronym FOMO (fear of missing out)[*] first appeared in 2004, after the global conquest of social media. Maybe computer technology will also be the solution – AI chat bots can generate a reasonably convincing conversation, but it feels on the shallow end at the moment. One of my colleagues was concerned that after using it as a cheap form of therapy, he'd somehow corrupted the core AI programme, because, like Marvin the Paranoid Android, his life was too depressing. AI conversation isn't real connection, it just predicts answers to questions. There doesn't appear to be a substitute for in-person connection – except maybe pets, ownership of which can reduce social isolation.[13] Still I'd rather chat to a mate than a dog; most of my friends don't eat fox faeces or lick their own balls.

---

[*] Not to mention FOJI (fear of joining in), JOMO (the joy of missing out), FOBO (fear of better options) and the inexplicable FOMOMO (Fear Of the Mystery Of Missing Out), a term so millennial in its origins that I struggle to understand it.

SOCIAL ANIMALS

How does any of this work? The interconnectivity of human health means biological, psychological and behavioural processes all contribute to disease, and social isolation influences all of these aspects. It's a truism to explain the impact of isolation by stating that humans are 'social animals'; most animals are social, even some bacteria! But humans need social co-operation to survive, and this has almost certainly been an important part of our evolution. The need for company is hard-wired. Recent research demonstrating that craving for social contact fires the substantia nigra, the region of the brain associated with hunger.[14] The downsides of loneliness return us to our favourite signalling molecule – cortisol. Isolation elevates cortisol, it even has a delayed action, with loneliness the day before translating to more cortisol the following morning;[15] FOMO causes physical consequences! Elevated cortisol in response to loneliness causes the same negative impacts as sustained stress, and its interplay with inflammation is a recurring theme in bad outcomes. The inverse has been observed: happiness lowers cortisol, in British civil servants at least.[16] Isolation also impacts the psychology and behaviours of the lonely person – when stressed and alone, you might conceivably turn to drinking and smoking, possibly to replace the dopamine hit of company. This clearly has biological mechanisms – they have yet to be ascertained. There will be a reason for everything, a chemical reaction in a cell somewhere in your body that links dancing to dementia. That we don't know what that happens to be (yet) is cause for wonderment. We will never know all the answers, and that's OK. It also means that health cannot be simplified to a single thing someone can monetise.

THE FINAL EXPERIMENT

As my final experiment I decided to isolate for a week whilst I wrote this final chapter of the book. It was the first occasion I had truly been on my own for a sustained time period in nearly twenty-five years. The last solitary spell had been when I took ten days to deliver a car from New York to San Diego and slowly lost my grip on reality; at one point I tested how far I could drive with my eyes shut (three seconds), with my hands off the wheel (ten minutes) and following another specific driver at one car's length (one hour). I also found myself trying to eat a seventy-two-ounce steak having seen an advert for it 350 miles previously (reader I couldn't). This time was not like that – but I felt quite nervous going into it. Would I hate it, or worse, would I like it too much? I must also admit that it wasn't under the worst circumstances, as I write these words overlooking the sea in Cornwall on a crystal-clear January morning. I can report no immediate negative health effects: I did more exercise, I ate fewer fruits and vegetables, I drank a bit more wine, I slept more, I didn't swim in the sea (it was January and my cold tolerance only goes so far), I grew a beard and didn't wash my hair, I loaded the dishwasher exactly the right way. Cooking a whole roast chicken for one was a low point. I didn't notice any significant increase in stress, but being able to see the sea and not having children asking how to install apps on their phones or colleagues demanding that I complete risk assessments might have balanced out the loneliness. And it was not true social isolation; whilst I spoke to only three people in the flesh, the publican, the shopkeeper and a lady whose dog hated me (twice, in the same place on separate days), I spent a lot of the time on social media, particularly WhatsApp, chatting to my family. I also had two video calls and took part in an online conference. But isolation isn't being on your own *per se*, it is about being alone. Being on my own, but with a purpose, meant I didn't notice the lack of family. The time pressure of getting the book written distracted me from the loneliness; when I finally finished, then I noticed that I was all alone.

Allowing me to add one more line to my results table

Intervention	Why	Impact
A week alone	Experience isolation	Sense of purpose, longer beard

Table 4: The final revelation of Professor John. A bonus reflection.

Having a purpose is central to happiness. Work by Professor Laura Carstensen at Stanford shows that our sense of purpose changes throughout our lives. When younger and our time feels endless, we prioritise expanding our horizons and making new friends. As we enter the later acts of our life's play, we prioritise connections with the people we already know and laying down our legacy. Her work also shows the fluidity of these priorities; in times of crisis, for example pandemics, we circle the wagons and focus on our nearest and dearest. As Atul Gawande said in his commencement speech at Stanford in 2021: 'As you get older you begin to know yourself – your capabilities, your gaps, what motivates you – well enough to commit to efforts that can take a long time to realize.'[17] The limits of our lives become more obvious, reducing the paralysis of choice. When young, we feel we can do anything and worry about missing out; when more fully formed, we know what we enjoy and can focus on what we want to achieve with our remaining allocation of time.

Discovering my purpose on a rocky headland overlooking the sea was great for me, but of all the 'experiments' that I undertook, it was the least transferable to anyone else. Let's return to the more population-level research as we near our end.

OUT OF THE BLUE

Is there any further evidence that combining a healthy diet with moderate exercise and good social networks can extend your life? Well, yes, there is.* Some regions of the world have a greater than expected proportion of nonagenarians and centenarians (ninety- and hundred-year-olds). The statisticians who identified these areas, Michel Poulain and Giovanni Pes,[18] named them blue zones, apparently after the colour they used on the map to indicate them. The term really gained popular credence following a 2005 *National Geographic* article by Dan Buettner.[19] It has of course now got a trademark. The areas include Sardinia (the original from Poulain's study), Okinawa, Loma Linda (California), the Nicoya Peninsula in Costa Rica, and Icaria, Greece. Some of the overlapping characteristics of people living in these areas include moderate activity, social engagement and healthy eating. This suggests that combining a healthy lifestyle with social connectivity can indeed extend

---

* Of course there is, it'd be a ludicrous way to end a book, posing a rhetorical question and then just saying no. Unless the question was can we Live Forever?

the healthspan. An alternative view though is that these simply represent upper outliers on the bell curve of human longevity, so for every Okinawa (life expectancy eighty-four years) there is a Lesotho (life expectancy fifty).

But one striking thing shared with at least three of the five zones is that elderly individuals living there, who would have been born in the early twentieth century, went through some extreme events in their early lives. Okinawa was the site of a brutal eighty-day campaign in 1945 when much of the island was flattened; during Axis occupation in the period 1941–4, Greece experienced a famine contributing to 300,000 deaths and Italians consumed the lowest numbers of calories (about one thousand day) of nearly any European country during the Second World War. This suggests that trauma, or at least calorie restriction, can have long-term health benefits.

A similar effect occurred in Cuba during the Special Period (the time following the collapse of the USSR, their main trading partner). During this period, daily calorie intake fell from three thousand (too much) to two thousand (too little, given the increase in physical exercise with people getting to work by bicycle or walking). Accordingly, obesity and the rate of type 2 diabetes decreased,[20] whilst life expectancy increased. Putting your population through a period of famine isn't great governmental policy – there are countless examples of where famine causes both short-term mortality and long-term effects. The Dutch Hunger Winter (1944–5) had permanent, negative effects on the babies in the womb and – remarkably – their children too![21]

Extremes of calorific intake in childhood have lasting consequences – obesity as much as starvation. And this leads us to my penultimate suggestion . . .

THE HEALTH PENSION

This advice is particularly for the readership in their late teens/early twenties, who still have time to change things. If you are older, the best thing you can do is get a hobby and find a group of like-minded people with whom you properly connect. It could be a veterans' association (though admittedly that does necessitate being in the forces in the first place), a running club, an allotment society or a church.[*]

However, if you are younger, there's still time. Many of our organs have a linear decay from their peak in our mid-twenties. This is seen for lung function, brain connectivity, cardiac output, kidney function, hearing. Some of these we can't actively improve and need just to protect: kidneys, ears, livers. But some we can train and improve, giving us more to lose. The higher the levels, the longer we can delay decay, particularly the brain, but also the heart and lungs. Educational attainment protects against dementia; analysis of nuns' childhood diaries for complexity of phrasing predicted dementia outcomes sixty years later.[22] This is where The Health Pension™ comes in.[†] The best time to pay into a cash pension is as early as possible in your life, giving the money much longer to accumulate, especially from compound interest. Likewise, in my Health Pension™, if you can raise the function of your organs in your youth, you have more you can draw upon later in life. Get out there, read books, learn useless facts, take higher degrees.

---

[*] But not a cycling chain-gang; no one needs any more MAMILs (Middle Aged Men in Lycra) blocking traffic on country roads, thank you.
[†] Not actually a trademark, but you're no one if you're not monetising your findings.

Likewise, do as much exercise as you can, push your lung function to the limit. But critically, be as happy as you can, approach every problem with a smile. It all sounds trite, but there's a good chance it will pay off later. A slight caveat comes from the former DCMO Sir JVT: playing football in your twenties and then doing nothing for the next twenty years isn't going to help that much. Exercise needs to be a lifelong commitment.

Admittedly, the Health Pension™ may not be the most original idea – Lewis Carroll was very much onto the same thing:

'You are old,' said the youth, 'and your jaws are too weak
For anything tougher than suet;
Yet you finished the goose, with the bones and the beak—
Pray, how did you manage to do it?'

'In my youth,' said his father, 'I took to the law,
And argued each case with my wife;
And the muscular strength, which it gave to my jaw,
Has lasted the rest of my life.'

Lewis Carroll, 'You Are Old, Father
William', *Alice's Adventures in Wonderland*

Sunlit Uplands

I am (hopefully) entering only the third of four quarters of my life. I don't really want to know what it would feel like if I were entering the home stretch, staring down the barrel of social care and total cognitive failure, but I'll leave that for sixty-year-old me to worry about. As Atul Gawande says, 'We only die once';[1] and no one really knows what the end-game will entail. Time is fleeting, and as fast as I run to keep up with the sun it will still sink.

I started this journey seeing grey hairs in the mirror and worrying about my impending death. One year later, with considerably *more* grey hairs, what did I learn by metaphorically peeling back my skin to look at my organs? I came to realise that I'm less worried about death than ageing and my inevitable decline. Making some, not unreasonable, changes now doesn't seem too big a price to pay.

I thought my journey of discovery might change my diet or exercise regime, and I tried to do this. It never stuck. What did change was that I discovered a profound sense of gratitude for the time I spend with my family and friends, just being together. And I'd recommend you do the same. My final piece of advice is: 'Do more of what you like, with people you love.' Live your life in full technicolour through the connection with others and then – when the time comes for it all to fade to black – maybe it'll just take a bit longer for those colours to ebb away. In the end, our lives aren't about not dying or not ageing, but accepting the path. This contains the essence of ageing well.

As a seventeen-year-old Hunter S. Thompson (my literary hero, but not necessarily a role model for healthy living) wrote in an article entitled 'Security':

'Who is the happier man, he who has braved the storm of life and lived or he who has stayed securely on shore and merely existed?'[2]

Glossary

Adrenaline. Fast acting fright, flight or fight hormone.

Aetiology. Science word for the cause or causes of a disease; sounds vaguely like a bookish Viking king.

Allele. One of the multiple possible variants of each individual gene. Our alleles make us individual.

Alzheimer's disease. A type of dementia characterised by an accumulation of a protein called Aβ amyloid in the brain. (Not all dementia is Alzheimer's.)

Antibody. Produced by the immune system to bind to and destroy pathogens; they recognise tiny structural motifs in other biological molecules.

Arrythmia. The heart beating out of time with itself.

Atherosclerosis. When fat, cells and other rubbish block up the blood vessels.

Bauxite. Aluminium ore (and town in USA).

Blinding. In clinical trials where participants (and often the people running the study) don't know who received what.

Blueberries. Whilst tasty, often mis-sold as a superfood to cure all that ails you. To such an extent, that they (to me) are a shorthand for lazy wellness hacks. Less is made about their country cousins, the bil-, blae- or whortleberry, let alone the *myrtilles sauvages* for the Francophile. They likely have more benefit through the joy of foraging with friends on a late summer's day on the Yorkshire Dales (ideally saying 'boop' when you find one, don't ask). NB only eat wild berries if you are absolutely sure what you are doing! Some blue berries in the wild will for sure shorten your life.

BMI. Body mass index. A slightly crude way of comparing weight normalised for height. It is calculated as height/mass^2 (in metric).

Bronchodilator. Drug that opens the airways, e.g. Ventolin.

Cancer. When cells grow out of control and start to invade other tissues.

CAR-T. Re-engineered immune cell that recognises and attacks cancer cells.

Cholesterol. A good news/bad news fat molecule. It makes your cell membranes, but also kills you if you have too much.

Chromosome. DNA's higher order organisation. How your genes are packaged together. Each cell contains twenty-three pairs (making forty-six). You inherit one of each pair from each parent. The twenty-third pair (the sex chromosome) defines biological sex: XX is female, XY male (there are other combinations causing a range of hormonal conditions in about 1.7% of people, whom some scientists consider to be intersex – making it about as common as red hair).

Clinical trial. A study of an intervention in people. Ideally blinded and placebo-controlled. Groups in clinical studies are often called arms; this can be active or control.

Cohort. A group of people in a clinical study, or more generally just a group of similar people.

Communicable disease. One passed from one person to another – also known as infectious disease, expertly covered in the book *Infectious* (available from good stockists everywhere).

COPD. Chronic obstructive pulmonary disease – progressive destruction of the lung, reducing the ability to breathe.

Correlate. A surrogate measure in a scientific study with a similar pattern to the disease. Not the same as causation. Remember all people who confuse correlation with causation will die.

Cortisol. Slow-acting fright, flight or fight hormone. Key component of stress.

Dementia. The progressive loss of connectivity in the brain, leading to loss of brain function – often first characterised by loss of memory. Alzheimer's disease is a subset of dementia.

Diabetes. Too much glucose in the blood. Has two types. Type 1 (T1D) is an autoimmune condition associated with the loss of the cells that produce insulin. Type 2 (T2D) is characterised by insulin resistance.

DNA. Curly wurly molecule of inheritance.

Dominant (genetics). Of the gene allele that presents as the phenotype – the classic, but subtly wrong, example is brown eye colour.

ECG. Electrocardiogram, measures the heart's electrical current.

Epigenetic. A system by which cells instruct themselves how much protein to make from their DNA. It's a quick reference system, a bit like bookmarks, and affects gene accessibility. The word *epi* means 'outside of'; so epi-genetic, outside the genes.

Erythrocyte. Red blood cells, carriers of oxygen and carbon dioxide.

F1 – for Filial generation 1. Often found on seed packets. F1 crosses are more vigorous than either purebred parental plant, but they don't breed true, ensuring you return to buy new seeds next year (a pansy scheme if you will).

FEV. Forced expiratory volume, how much air you can blow down a tube in a second.

FVC. Forced vital capacity. The total amount of air you can breathe out.

Genes. The unit of inheritance, found on DNA. Each one encodes a single protein.

Genetics. The study of genes. Coined by William Bateson (from the Greek word meaning 'origin') in 1905. Curiously the word 'genetics' predates the word 'gene', first used in 1909 by Wilhelm Johannsen.

Genotype. The genetic make-up of a person, animal, plant of bacteria.

Goldilocks effect. The just right choice – from the fairy tale character who, whilst engaging in a study of ursine behaviour with relation to bedding, seating and eating, chose the median of three options.

Hand-wavy. A (relatively) polite way of saying someone's theory is wafty bullshit. It comes from the hand-waving associated with trying to explain an unsubstantiated idea.

Heart attack. Also referred to as myocardial infarction, or just 'a coronary'. The blood supply to the heart stops, the heart dies and the body dies soon after.

Heart failure. The heart is still working but not pumping enough blood to support the rest of the body.

Homeostasis. The body's regulation of its physical and chemical conditions, keeping them more or less constant.

Hyper-. Greek stem meaning 'too high'; seen in hyperglycaemia (too much sugar), hyperthermic (too hot) and hypertension (too high blood pressure). Opposite is hypo-.

Hypo-. Greek stem meaning 'too low'; seen in hypoglycaemia (too little sugar), hypothermic (too little heat) and hypotension (blood pressure too low). Opposite is hyper-

Hypothesis. A testable idea. If you are Team Popper, it is falsifiable too.

Idiopathic. Without known cause; what medics call something when they have no idea but don't want to admit it.

In vitro, in vivo, in silico. Different types of science experiment. *In vitro* usually uses cells in petri dishes (vitro is Latin for glass). *In vivo* uses animals. *In silico* uses computers – from silicon chips.

Inflammation. Boo hiss, the villain of the piece (except when it isn't). Your body's natural reaction to an infection, and gets cells to the right place. Problematic when it doesn't resolve.

Inhaled corticosteroids (ICS). Drugs that mimic the hormone cortisol that are delivered by an inhaler (normally brown in colour) and reduce disease in asthma (and COPD).

Low T. Reduced testosterone; probably made up by men who inexplicably feel they are missing out on the menopause.

Lucy. A 3.2 million year old, somewhat complete, fossil skeleton found in Ethiopia; one of the earliest proto-humans, technically an australopithecine. Named after the song 'Lucy in the Sky with Diamonds'.

Mens sana in corpore sano. Favourite aphorism of nineteenth-century educationalists – 'a healthy mind in a healthy body'.

Meta. Combining several studies increases the accuracy because some of the individual biases drop out; from the Greek for 'beyond'.

Microbiome. The bacteria (and other microorganisms) that live on, in and around our body. Extraordinarily diverse, complex ecosystem of microscopic life. Correlates with a lot of stuff. Who knows if it causes anything?

Mitochondria. The powerhouse of the cell, burns glucose in oxygen to create energy for the cell.

Monogenic. A single gene determining a single trait.

Moore's Law. A doubling of computing power every two years; when applied to sequencing it would be a halving of price every two years.

Nepo Baby. Short for nepotism baby, someone who has their position through dint of parentage, not necessarily talent. Choose your own examples.

Neurotransmitter. Chemical substance by which nerve cells talk to each other.

Nominative determinism. When your name influences your work, blacksmiths named Smith for example. Also Mr Tickle, who, though he lacked a formal job, did do exactly what his name says.

Non-communicable disease. Diseases you can't catch from other people – people use this term for the majority of ageing diseases described in this book. Just to be slightly confusing, there may be an infectious trigger for a lot of these conditions.

Normal distribution. For a continuous variable (like height), people tend to cluster around a mean value, with more in the middle and fewer at the extremes. The curve is bell shaped.

Oncogene. A gene that in a healthy cell doesn't cause harm but has the capacity to cause cancer when mutated.

Phenotype. The physical manifestation of our genes.

Placebo. The control treatment, looks just like the real thing, but has no active ingredients.

Placebo effect. Where the act of taking part in the study alone improves outcomes measured.

Plasma. The wet bit of blood.

Pleiotropic. Multi-function (an admission by biologists that we don't really know what a compound does), see also Idiopathic.

PM2.5. A cocktail of particulate material less than 2.5 microns in size that you don't want to be inhaling.

Polygenic. Of genes, when more than one gene leads to a trait (this accounts for most attributes from eye colour to intelligence).

Popperian/Baconian epistemology. A way of doing science – ideas need to be tested and the knowledge that supports them can only come from observation. After Francis Bacon and Karl Popper.

PT-109. The Patrol torpedo boat commanded by Lt John Fitzgerald Kennedy. Sliced in half by a Japanese destroyer, Kennedy towed another crewman ashore, swimming three and a half miles with the life jacket strap clenched between his teeth.

PTI. Physical Training Instructor. A role in the army occupied by sadistic narcissists with a limited number of 'hilarious' catchphrases, including: 'Twice around my beautiful body, Go!'

Recessive. (of genetics) the silent partner gene allele. The opposite of dominant.

Ribosome. Tiny protein factory found inside the cell. It takes the message from the RNA and translates it into protein.

RNA. Messenger molecule from DNA to protein.

SCI unit. Science communication unit – Wales for area, swimming pool for volume, whales for weight (confusingly).

Sensitivity. Ability to identify a positive case.

SI units. Standardised unit for weights and measures.

SNP. Single nucleotide polymorphisms – point mutations (changes in a single base pair) in a gene/allele. They can be coding (in the gene itself) or non-coding (in the bit of DNA surrounding the gene).

Specificity. Ability to identify a negative case.

Standard deviation (or SD). How varied a population is around a central average (mean) value.

Telomere. The ends of the DNA chromosome. Loved by anti-ageing fanatics seeking eternal life, no longer in fashion with actual scientists.

Thrombosis. Fancy word for blood clot.

Tumour. The cancerous cell/cells.

ULEZ. Ultra Low Emission Zone. A clever way to reduce pollution and infuriate the antis at the same time.

UV. Light at the blue, destructive end of the spectrum. Gives you skin cancer, to be avoided.

White Coat Syndrome. Elevated blood pressure measurement in a doctor's surgery caused by the act of having your blood pressure measured.

Acknowledgements

Having confidently told my wife that I was never going to write another book, I lay the blame for this one firmly at the door of my fabulous agent Caroline Hardman, who (as before) sparked the initial idea over a cup of tea.* For this I am of course eternally grateful, though my wife less so!

Thanks of course to the team at Oneworld: my editor Sam Carter for entering the breach of my terrible grammar and punctuation once more; Hannah Haseloff for dealing admirably with my chaotic process; Sam Wells for sympathetic editorial improvements; and Juliet and Novin for founding and being the best of all publishers.

My family were extraordinarily tolerant of me disappearing into the study for a significant proportion of every weekend, even allowing me a week away in Cornwall to cosplay being a writer. They also tolerated me when I starved myself or overloaded on fibre – true love. And my extended family for me testing levels of science literacy.

Whilst I am exceptionally wise, I supplemented this wisdom by speaking to a huge cohort of helpful and

---

* Thanks too to Thomas Sullivan and Charles Edward Taylor: inventor of the teabag and founder of Yorkshire Tea respectively. I consumed a lot of tea writing this; sadly the anti-ageing effects are pretty marginal.

supportive people who were astonishingly generous with their time, putting up with my daft questions. In no particular order this includes: Cathy Slack, Luke O'Neill, Peter Barnes, Mike Wall, my parents and sisters, Gary Frost, Bill Wisden, Nick Oliver, Julian Marchesi, James Kinross, with a special shout out to Despoina Chrysostomou for running analysis on unmentionable samples, David Thomas, David Nutt, JVT, Sian Henson, Neil Hill, Adam Rutherford, Rachel Bastiaenen, Andrew the ECG technician, Eva Fiorenzo, Hugo Farne and Gary Fuller. I am grateful to my colleagues at Imperial who took the time to read and check for silly mistakes: Jo Jackson, Jesus Rodriguez-Manzano, Laki Buluwela and Vanessa Sancho Shimizu. Thanks also to my friends on the Signal group 'pedantry corner' who think they know more better grammar than me (or is that I?). I am of course grateful for the time to write and support from Imperial, particularly from Wendy Barclay and Robin Shattock. And thanks to my research group for putting up with my endless chat about ageing, microbiomes and especially for not quitting when I was hangry for a week on the zero-calorie diet.

Thanks to Ben Willbond for advice about how to structure a narrative and Al Foster for encouragement to self-experiment. Finally, Al, Jim, Tony and the rest of the We Have Ways team for letting me waffle on about health in war. Thank you to Frank Turner for letting me use some of his amazing lyrics.

As before, the illustrations are by the wonderfully talented Ash Uruchurtu and supported by a Communicating Immunology grant from the British Society for Immunology.

References

PROLOGUE: TIME AND TIDE

1 Tregoning, J.S., *Infectious: Pathogens and how we fight them.* 2021, London: Oneworld.

2 WHO, *Urgent action needed to tackle stalled progress on health-related Sustainable Development Goals.* 2023; Available from: https://www.who.int/news/item/19-05-2023-urgent-action-needed-to-tackle-stalled-progress-on-health-related-sustainable-development-goals.

3 ONS, *Death registration summary statistics, England and Wales: 2022.* 2023; Available from: https://www.ons.gov.uk/peoplepopulationandcommunity/birthsdeathsandmarriages/deaths/articles/deathregistrationsummarystatisticsenglandandwales/2022.

4 Horiuchi, A. and Y. Nakayama, 'Colonoscopy in the sitting position: lessons learned from self-colonoscopy by using a small-caliber, variable-stiffness colonoscope.' *Gastrointestinal Endoscopy,* 2006. 63(1): pp. 119–20.

5 Smith, M.L., 'Honey bee sting pain index by body location.' *PeerJ,* 2014. 2: p. e338.

1 LET US END AT THE BEGINNING

1 Shemie, S.D., et al., 'International guideline development for the determination of death.' *Intensive Care Med,* 2014. 40(6): pp. 788–97.

2 Bar-On, Y.M., R. Phillips, and R. Milo, 'The biomass distribution on Earth.' *Proceedings of the National Academy of Sciences,* 2018. 115(25): pp. 6506–11.

3 López-Otín, C., et al., 'The hallmarks of aging.' *Cell,* 2013. 153(6): pp. 1194–217.

4 Fick, Laura J., et al., 'Telomere Length Correlates with Life Span of Dog Breeds.' *Cell Reports*, 2012. 2(6): pp. 1530–6.

5 Miller Jr, W.B., and J.S. Torday, 'Reappraising the exteriorization of the mammalian testes through evolutionary physiology.' *Commun Integr Biol*, 2019. 12(1): pp. 38–54.

6 Attarwala, H., 'TGN1412: From Discovery to Disaster.' *J Young Pharm*, 2010. 2(3): pp. 332–6.

7 Kaptchuk, T.J., et al., "Placebos without deception: a randomized controlled trial in irritable bowel syndrome.' *PLoS One*, 2010. 5(12): p. e15591.

8 Robert, W.Y., et al., 'Parachute use to prevent death and major trauma when jumping from aircraft: randomized controlled trial.' *BMJ*, 2018. 363: p. k5094.

9 Rosenblueth, A. and N. Wiener, 'The role of models in science.' *Philosophy of Science*, 1945. 12(4): pp. 316–21.

10 Tsitsimpikou, C., et al., 'Dietary supplementation with tomato-juice in patients with metabolic syndrome: a suggestion to alleviate detrimental clinical factors.' *Food Chem Toxicol*, 2014. 74: pp. 9–13.

11 Cheng, H.M., et al., 'Lycopene and tomato and risk of cardiovascular diseases: A systematic review and meta-analysis of epidemiological evidence.' *Critical Reviews in Food Science and Nutrition*, 2019. 59(1): pp. 141–58.

12 Mencken, H.L., 'The Divine Afflatus', in Mencken, H.L., *Prejudices: Second Series*. 1920, New York: Alfred A. Knopf; pp. 155–71, p. 158.

2 THE FAULT IN OUR GENES

1 Rutherford, A., *Control: The Dark History and Troubling Present of Eugenics*. 2023, London: Weidenfeld and Nicholson.

2 Markt, S.C., et al., 'Sniffing out significant "Pee values": genome wide association study of asparagus anosmia.' *BMJ*, 2016. 355: p. i6071.

3 Eiberg, H., et al., 'Blue eye color in humans may be caused by a perfectly associated founder mutation in a regulatory element located within the HERC2 gene inhibiting OCA2 expression.' *Human Genetics*, 2008. 123(2): pp. 177–87.

4 Chrisafis, A., 'Rise in domestic mishaps puts strain on NHS.' *Guardian*, 2001.

5 Nafilyan, V., et al., 'Risk of death following COVID-19 vaccination or positive SARS-CoV-2 test in young people in England.' *Nature Communications*, 2023. 14(1): p. 1541.

6 Marshall, G., et al., 'Streptomycin treatment of pulmonary tuberculosis.' *Br Med J*, 1948. 2(4582): pp. 769–82.

7 Doll, R., 'Experiences of a battalion medical officer in the retreat to Dunkirk: I.' *BMJ*, 1990. 300(6733): pp. 1183–6.

8 Doll, R. and A.B. Hill, 'The mortality of doctors in relation to their smoking habits; a preliminary report.' *BMJ*, 1954. 1(4877): pp. 1451–5.

9 Kashima, Y., et al., 'Single-cell sequencing techniques from individual to multiomics analyses.' *Experimental & Molecular Medicine*, 2020. 52(9): pp. 1,419–27.

10 Jorde, L.B. and M.J. Bamshad, 'Genetic Ancestry Testing: What Is It and Why Is It Important?' *JAMA*, 2020. 323(11): pp. 1089–90.

11 Agence France-Presse in The Hague, 'Dutch fertility doctor "secretly fathered at least 49 children".' *Guardian*, 2019.

12 Zerjal, T., et al., 'The Genetic Legacy of the Mongols.' *The American Journal of Human Genetics*, 2003. 72(3): pp. 717–21.

13 23&me, *23andMe And The FDA*; Available from: https://customercare.23andme.com/hc/en-us/articles/211831908-23andMe-and-the-FDA.

14 Herper, M., *23andMe Gets $300 Million Boost From GlaxoSmithKline To Develop New Drugs*. 2018; Available from: https://www.forbes.com/sites/matthewherper/2018/07/25/23andme-gets-300-million-boost-from-glaxo-to-develop-new-drugs/.

15 Siddiqui, Z., '23andMe notifies customers of data breach into its "DNA Relatives" feature.' *Reuters*, 2023.

16 Beier, J.I. and G.E. Arteel, 'Environmental exposure as a risk-modifying factor in liver diseases: Knowns and unknowns.' *Acta Pharmaceutica Sinica B*, 2021. 11(12): pp. 3768–78.

3 A PICTURE OF HEALTH

1 Statista, *In Vitro Diagnostics – Worldwide*. 2023; Available from: https://www.statista.com/outlook/hmo/medical-technology/in-vitro-diagnostics/worldwide.

2 Emerging Risk Factors Collaboration, 'Diabetes mellitus, fasting blood glucose concentration, and risk of vascular disease: a collaborative meta-analysis of 102 prospective studies.' *Lancet*, 2010. 375(9733): pp. 2,215–22.

3 Criado-Perez, C., *Invisible women : data bias in a world designed for men*. 2020, New York: Abrams.

4 Patel, R., et al., 'Evaluation of the uptake and delivery of the NHS Health Check programme in England, using primary care data from 9.5 million people: a cross-sectional study.' *BMJ Open*, 2020. 10(11): p. e042963.

5 Wilkinson, E., 'The rise of direct-to-consumer testing: is the NHS paying the price?' *BMJ*, 2022. 379: p. o2518.

6 Student, 'The Probable Error of a Mean.' *Biometrika*, 1908. 6(1): pp. 1–25.

7 Galton, F., 'The Ballot-Box.' *Nature*, 1907. 75(1952): p. 509.

8 Brett, L., *PMSL: Or How I Literally Pissed Myself Laughing and Survived the Last Taboo to Tell the Tale*. 2020, London: Green Tree.

4 THE HEART OF THE MATTER

1 Zhang, D., W. Wang, and F. Li, 'Association between resting heart rate and coronary artery disease, stroke, sudden death and noncardiovascular diseases: a meta-analysis.' *CMAJ*, 2016. 188(15): pp. e384–92.

2 Mujika, I., 'The cycling physiology of Miguel Indurain 14 years after retirement.' *Int J Sports Physiol Perform*, 2012. 7(4): pp. 397–400.

3 Hailu, R., *Fitbits and other wearables may not accurately track heart rates in people of color*. 2019; Available from https://www.statnews.com/2019/07/24/fitbit-accuracy-dark-skin/.

4 Bergmann, O., et al., 'Evidence for cardiomyocyte renewal in humans.' *Science*, 2009. 324(5923): pp. 98–102.

5 The King's Fund, *Key facts and figures about the NHS*. 2023; Available from: https://www.kingsfund.org.uk/audio-video/key-facts-figures-nhs.

6 Public Health England, *Health matters: preventing cardiovascular disease*. 2019; Available from: https://www.gov.uk/government/publications/health-matters-preventing-cardiovascular-disease/health-matters-preventing-cardiovascular-disease#cvd-ambitions-and-secondary-prevention.

7 Dodge Jr, J.T., et al., 'Lumen diameter of normal human coronary arteries. Influence of age, sex, anatomic variation, and left ventricular hypertrophy or dilation.' *Circulation*, 1992. 86(1): pp. 232–46.

8 Galkina, E. and K. Ley, 'Immune and inflammatory mechanisms of atherosclerosis.' *Annu Rev Immunol*, 2009. 27: pp. 165–97.

9 Martin, P. *Is it normal to break ribs during CPR?* 2022; Available from: https://www.procpr.org/blog/training/hear-ribs-break-cpr.

10 Mensah, G.A., et al., 'Decline in Cardiovascular Mortality: Possible Causes and Implications.' *Circ Res*, 2017. 120(2): pp. 366–80.

11 Williams, D.H., *Stab Wound of the Heart and Pericardium – Suture of the Pericardium – Recovery –Patient Alive Three Years Afterward*. 1897, New York: Publishers' Printing Company.

12 Sommerhaug, R.G., et al., 'Multiple (more than eight) bypass grafts in severe diffuse coronary disease: improved exercise tolerance and functional classification in seventy-seven consecutive patients.' *Am Heart J*, 1987. 114(4 Pt 1): pp. 710–7.

13 Stamatakis, E., et al., 'Association of wearable device-measured vigorous intermittent lifestyle physical activity with mortality.' *Nature Medicine*, 2022. 28(12): pp. 2521–9.

5 A STROKE OF BAD LUCK

1 Johnson A.B., *Hemorrhage*. 2023, Treasure Island, Florida: Stat Pearls.

2 Sender, R., et al., 'The total mass, number, and distribution of immune cells in the human body.' *Proceedings of the National Academy of Sciences*, 2023. 120(44): p. e2308511120.

3 Hatton, I.A., et al., 'The human cell count and size distribution.' *Proceedings of the National Academy of Sciences*, 2023. 120(39): p. e2303077120.

4 Ribatti, D. and E. Crivellato, 'Giulio Bizzozero and the discovery of platelets.' *Leukemia Research*, 2007. 31(10): pp. 1,339–41.

5 Josefsson, E.C., W. Vainchenker and C. James, 'Regulation of Platelet Production and Life Span: Role of Bcl-xL and Potential Implications for Human Platelet Diseases.' *Int J Mol Sci*, 2020. 21(20).

6 Turetz, M., et al., 'Epidemiology, Pathophysiology, and Natural History of Pulmonary Embolism.' Semin Intervent Radiol, 2018. 35(2): pp. 92–8.

7 CDC, *Stroke Facts*. 2024; Available from: https://www.cdc.gov/stroke/data-research/facts-stats/?CDC_AAref_Val=https://www.cdc.gov/stroke/facts.htm.

8 Eichinger, S., et al., 'Venous thromboembolism in women: a specific reproductive health risk.' *Human Reproduction Update*, 2013. 19(5): pp. 471–82.

9 Grillo, A., et al., 'Sodium Intake and Hypertension.' *Nutrients*, 2019. 11(9).

10 He, F.J., J. Li, and G.A. Macgregor, 'Effect of longer term modest salt reduction on blood pressure: Cochrane systematic review and meta-analysis of randomised trials.' *BMJ*, 2013. 346: p. f1325.

11 Mente, A., et al., 'Urinary sodium excretion, blood pressure, cardiovascular disease, and mortality: a community-level prospective epidemiological cohort study.' *Lancet*, 2018. 392(10146): pp. 496–506.

12 do Amaral, M.A.S., et al., 'Effect of music therapy on blood pressure of individuals with hypertension: A systematic review and Meta-analysis.' *International Journal of Cardiology*, 2016. 214: pp. 461–4.

13 Kapil, V., et al., 'Dietary nitrate provides sustained blood pressure lowering in hypertensive patients: a randomized, phase 2, double-blind, placebo-controlled study.' *Hypertension*, 2015. 65(2): pp. 320–7.

14 WHO, *The use of stems in the selection of International Nonproprietary Names (INN) for pharmaceutical substances 2018*. 2018, Geneva: World Health Organization.

15 Brunner, F.J., et al., 'Application of non-HDL cholesterol for population-based cardiovascular risk stratification: results from the Multinational Cardiovascular Risk Consortium.' *Lancet*, 2019. 394(10215): pp. 2173–83.

16 Esperland, D., L. de Weerd, and J.B. Mercer, 'Health effects of voluntary exposure to cold water – a continuing subject of debate.' *Int J Circumpolar Health*, 2022. 81(1): p. 2111789.

17 Knechtle, B., et al., 'Cold Water Swimming-Benefits and Risks: A Narrative Review.' *Int J Environ Res Public Health*, 2020. 17(23).

18 DHSC, *Sewage in water: a growing public health problem*. 2022; Available from: https://www.gov.uk/government/news/sewage-in-water-a-growing-public-health-problem.

19 Rew, K, *10 ways to stay well swimming*. 2022; Available from: https://www.outdoorswimmingsociety.com/10-ways-to-stay-well-swimming/.

20 Espeland, D., L. de Weerd, and J.B. Mercer, 'Health effects of voluntary exposure to cold water – a continuing subject of debate.' *International Journal of Circumpolar Health*, 2022. 81(1).

21 Zwaag, J., et al., 'The Effects of Cold Exposure Training and a Breathing Exercise on the Inflammatory Response in Humans: A Pilot Study.' *Psychosom Med*, 2022. 84(4): pp. 457–67.

6 A FATE WORSE THAN BREATH

1 University of Hertfordshire, *Research reveals what your sleeping position says about your relationship*. 2014; Available from: https://www.sciencedaily.com/releases/2014/04/140415203702.htm.

2 Robertson C.E., et al., 'Culture-Independent Analysis of Aerosol Microbiology in a Metropolitan Subway System.' *Applied and Environmental Microbiology*, 2013. 79(11): pp. 3,485–93.

3 Pottegård, A., et al., 'SearCh for humourIstic and Extravagant acro-Nyms and Thoroughly Inappropriate names For Important Clinical trials (SCIENTIFIC): qualitative and quantitative systematic study.' *BMJ*, 2014. 349: p. g7092.

4 De Sutter, A.I.M., L. Eriksson, and M.L. van Driel, 'Oral antihistamine-decongestant-analgesic combinations for the common cold.' *Cochrane Database of Systematic Reviews*, 2022(1).

5 Howes, D., 'Hiccups: a new explanation for the mysterious reflex.' *Bioessays*, 2012. 34(6): pp. 451–3.

6 Alvarez, J., et al., 'Evaluation of the Forced Inspiratory Suction and Swallow Tool to Stop Hiccups.' *JAMA Network Open*, 2021. 4(6): p. e2113933.

7 Davies, N., 'Drinking straw device is instant cure for hiccups, say scientists.' *Guardian*, 2021.

8 Farne, H., et al., 'Comparative Metabolomic Sampling of Upper and Lower Airways by Four Different Methods to Identify Biochemicals That May Support Bacterial Growth.' *Front Cell Infect Microbiol*, 2018. 8: p. 432.

9 Gibson, G.J., 'Spirometry: then and now.' *Breathe*, 2005. 1(3): p. 206.

10 Elizabeth, T.T., et al., 'Rate of normal lung function decline in ageing adults: a systematic review of prospective cohort studies.' *BMJ Open*, 2019. 9(6): p. e028150.

11 WHO, *WHO highlights huge scale of tobacco-related lung disease deaths*. 2019; Available from: https://www.who.int/news/item/29-05-2019-who-highlights-huge-scale-of-tobacco-related-lung-disease-deaths.

12 Font, A., et al., 'A tale of two cities: is air pollution improving in Paris and London?' *Environ Pollut*, 2019. 249: pp. 1–12.

13 Fuller, G., 'The evidence is clear: low-emission zones like London's Ulez work.' *Guardian*, 2023.

14 Christensen, L.T., C.D. Schmidt, and L. Robbins, 'Pigeon breeders' disease—a prevalence study and review.' *Clinical & Experimental Allergy*, 1975. 5(4): pp. 417–30.

15 Castillo, L. *Most Dangerous Sports Statistics*. 2023; Available from: https://gitnux.org/most-dangerous-sports-statistics/.

16 Wang, H., et al., 'Efficient Removal of Ultrafine Particles from Diesel Exhaust by Selected Tree Species: Implications for Roadside Planting for Improving the Quality of Urban Air.' *Environmental Science & Technology*, 2019. 53(12): pp. 6906–16.

17 Carlsen, K.H., et al., 'Exercise-induced asthma, respiratory and allergic disorders in elite athletes: epidemiology, mechanisms and diagnosis.' *Allergy*, 2008. 63(4): pp. 387–403.

18 ILC. Marathon or sprint: *Do elite-level athletes live longer than average?* 2023; Available from: https://ilcuk.org.uk/marathon-or-sprint/.

19 Lewis, A., et al., 'Singing for Lung Health-a systematic review of the literature and consensus statement.' *NPJ Prim Care Respir Med*, 2016. 26: p. 16080.

20 Holland, A.E., et al., 'Breathing exercises for chronic obstructive pulmonary disease.' *Cochrane Database of Systematic Reviews*, 2012(10).

7 THE SMOKING GUN

1 Baker, C. and Z. Mansfield, *Cancer statistics for England: Briefing Paper 06977.* 2023, London: House of Commons Library.

2 CRUK, *Cancer mortality for common cancers.* 2022; Available from: https://www.cancerresearchuk.org/health-professional/cancer-statistics/mortality/common-cancers-compared.

3 Harvard Medical School, *The science of sunscreen.* 2021; Available from: https://www.health.harvard.edu/staying-healthy/the-science-of-sunscreen.

4 Baker, R.R., E.D. Massey, and G. Smith, 'An overview of the effects of tobacco ingredients on smoke chemistry and toxicity.' *Food and Chemical Toxicology*, 2004. 42: pp. 53–83.

5 Daphne, C.W., M.E. Beverley, and J. Prabhat, 'Impact of vaping introduction on cigarette smoking in six jurisdictions with varied regulatory approaches to vaping: an interrupted time series analysis.' *BMJ Open*, 2022. 12(5): p. e058324.

6 David, T.L., et al., 'Examining the relationship of vaping to smoking initiation among US youth and young adults: a reality check.' *Tobacco Control*, 2019. 28(6): p. 629.

7 Salmon, C.P., M.G. Knize, and J.S. Felton, 'Effects of marinating on heterocyclic amine carcinogen formation in grilled chicken.' *Food and Chem Toxicology*, 1997. 35(5): pp. 433–41.

8 Sender, R. and R. Milo, 'The distribution of cellular turnover in the human body'. *Nature Medicine*, 2021. 27(1): pp. 45–8.

9 Falcaro, M., et al., 'The effects of the national HPV vaccination programme in England, UK, on cervical cancer and grade 3 cervical intraepithelial neoplasia incidence: a register-based observational study.' *Lancet*, 2021. 398(10316): pp. 2084–92.

10 Rider, J.R., et al., 'Ejaculation Frequency and Risk of Prostate Cancer: Updated Results with an Additional Decade of Follow-up.' *Eur Urol*, 2016. 70(6): pp. 974–82.

11 Mellie, R., Roger's *Profanisaurus: War and Piss*. 2018, London: Dennis.

12 Greten, F.R. and S.I. Grivennikov, 'Inflammation and Cancer: Triggers, Mechanisms, and Consequences.' *Immunity*, 2019. 51(1): pp. 27–41.

13 Krishnali, P., et al., 'Societal costs of chemotherapy in the UK: an incidence-based cost-of-illness model for early breast cancer.' *BMJ Open*, 2021. 11(1): p. e039412.

14 Laudicella, M., et al., 'Cost of care for cancer patients in England: evidence from population-based patient-level data.' *British Journal of Cancer*, 2016. 114(11): pp. 1286–92.

15 Hernandez, I., et al., 'Pricing of monoclonal antibody therapies: higher if used for cancer?' *Am J Manag Care*, 2018. 24(2): pp. 109–12.

8 A BRAIN IS ONLY AS STRONG AS ITS WEAKEST LINK

1 Wiehler, A., et al., 'A neuro-metabolic account of why daylong cognitive work alters the control of economic decisions.' *Current Biology*, 2022. 32(16): pp. 3564–75.e5.

2 Hardy, G.H., *A Mathematician's Apology*. 1940, Cambridge, Eng.: Cambridge University Press. pp. vii, 93, 1.

3 Murman, D.L., 'The Impact of Age on Cognition.' *Semin Hear*, 2015. 36(3): pp. 111–21.

4 William, S., 'Sport associated dementia.' *BMJ*, 2021. 372: p. n168.

5 Stewart, W. and K. Trujillo, *Modern Warfare Destroys Brains*. 2020, Cambridge, Mass.: Harvard Kennedy School.

6 Müller, U., P. Winter, and M.B. Graeber, 'A presenilin 1 mutation in the first case of Alzheimer's disease.' *Lancet Neurol*, 2013. 12(2): pp. 129–30.

7 Hampel, H., et al., 'The Amyloid-β Pathway in Alzheimer's Disease.' *Molecular Psychiatry*, 2021. 26(10): pp. 5481–503.

8 Sims, J.R., et al., 'Donanemab in Early Symptomatic Alzheimer Disease: The TRAILBLAZER-ALZ 2 Randomized Clinical Trial.' *JAMA*, 2023. 330(6): pp. 512–27.

9 Bui, C. and C. Miller, 'The Jobs You're Most Likely to inherit From Your Mother and Father.' *New York Times*, 2017.

10 Yan, R., et al., 'Synergistic neuroprotection by coffee components eicosanoyl-5-hydroxytryptamide and caffeine in models of Parkinson's disease and DLB.' *Proc Natl Acad Sci USA*, 2018. 115(51): pp. e12053–62.

11 Basurto-Islas, G., et al., 'Therapeutic benefits of a component of coffee in a rat model of Alzheimer's disease.' *Neurobiology of Aging*, 2014. 35(12): pp. 2701–12.

12 Owen, A.M., et al., 'Putting brain training to the test.' *Nature*, 2010. 465(7299): pp. 775–8.

13 Au, J., et al., 'There is no convincing evidence that working memory training is NOT effective: A reply to Melby-Lervåg and Hulme (2015).' *Psychonomic Bulletin & Review*, 2016. 23(1): pp. 331–7.

14 Jean François, D., et al., 'Playing board games, cognitive decline and dementia: a French population-based cohort study.' *BMJ Open*, 2013. 3(8): p. e002998.

15 Chételat, G., et al., 'Effect of an 18-Month Meditation Training on Regional Brain Volume and Perfusion in Older Adults: The Age-Well Randomized Clinical Trial.' *JAMA Neurology*, 2022. 79(11): pp. 1165–74.

16 Sparling, P.B., 'Legacy of Nutritionist Ancel Keys.' *Mayo Clinic Proceedings*, 2020. 95(3): pp. 615–7.

17 Pett, K.D., et al., 'The Seven Countries Study.' *European Heart Journal*, 2017. 38(42): pp. 3119–21.

18 OECD, *Health at a Glance*. 2023, Paris: OECD.

19 Bøggild, H. and A. Knutsson, 'Shift work, risk factors and cardiovascular disease.' *Scand J Work Environ Health*, 1999. 25(2): pp. 85-99.

20 National Toxicology Program, *NTP Review of Shift Work at Night, Light at Night, and Circadian Disruption*. 2021; Available from: https://ntp.niehs.nih.gov/whatwestudy/assessments/cancer/completed/shiftwork.

21 Li, J., M.V. Vitiello, and N.S. Gooneratne, 'Sleep in Normal Aging.' *Sleep Med Clin*, 2018. 13(1): pp. 1–11.

22 Otaiku, A.I., 'Distressing dreams in childhood and risk of cognitive impairment or Parkinson's disease in adulthood: a national birth cohort study.' *eClinicalMedicine*, 2023. 57: p. 101872.

23 Gentry, N.W., et al., 'Human circadian variations.' *J Clin Invest*, 2021. 131(16).

24 Milner, C.E. and K.A. Cote, 'Benefits of napping in healthy adults: impact of nap length, time of day, age, and experience with napping.' *J Sleep Res*, 2009. 18(2): pp. 272–81.

25 Sabia, S., et al., 'Association of sleep duration in middle and old age with incidence of dementia.' *Nat Commun*, 2021. 12(1): p. 2289.

26 Lin, F.R., et al., 'Hearing loss and incident dementia.' *Arch Neurol*, 2011. 68(2): pp. 214–20.

9 DON'T SUGAR-COAT IT

1 Lewis, G.F. and P.L. Brubaker, 'The discovery of insulin revisited: lessons for the modern era.' *J Clin Invest*, 2021. 131(1).

2 Conrad, N., et al., 'Incidence, prevalence, and co-occurrence of autoimmune disorders over time and by age, sex, and socioeconomic status: a population-based cohort study of 22 million individuals in the UK.' *Lancet*, 2023. 401(10391): pp. 18–90.

3 Ibid.

4 Klein, S.L. and K.L. Flanagan, 'Sex differences in immune responses.' *Nature Reviews Immunology*, 2016. 16(10): pp. 626–38.

5 Isaacs, S.R., et al., 'Viruses and Type 1 Diabetes: From Enteroviruses to the Virome.' *Microorganisms*, 2021. 9(7).

6 The Lancet Diabetes & Endocrinology (Editorial), 'COVID-19 and diabetes: a co-conspiracy?' *Lancet Diabetes & Endocrinology*, 2020. 8(10): p. 801.

7 Bjornevik, K., et al., 'Longitudinal analysis reveals high prevalence of Epstein-Barr virus associated with multiple sclerosis.' *Science*, 2022. 375(6578): pp. 296–301.

8 Lanz, T.V., et al., 'Clonally expanded B cells in multiple sclerosis bind EBV EBNA1 and GlialCAM.' *Nature*, 2022. 603(7900): pp. 321–7.

9 Warren-Gash, C., et al., 'Laboratory-confirmed respiratory infections as triggers for acute myocardial infarction and stroke: a self-controlled case series analysis of national linked datasets from Scotland.' *Eur Respir J*, 2018. 51(3).v.

10 Gardner, J., 'Two decades and $200 billion: AbbVie's Humira monopoly nears its end.' *BioPharmaDive*, 2023.

11 Department of Health, *NHS set to save £150 million by switching to new versions of most costly drug.* 2018; Available from: https://www.england.nhs.uk/2018/10/nhs-set-to-save-150-million-by-switching-to-new-versions-of-most-costly-drug/.

12 NHS, *Prescription Cost Analysis – England – 2021/22.* 2022; Available from: https://www.nhsbsa.nhs.uk/statistical-collections/prescription-cost-analysis-england/prescription-cost-analysis-england-202122.

13 Liang, G. and F.D. Bushman, 'The human virome: assembly, composition and host interactions.' *Nature Reviews Microbiology*, 2021. 19(8): pp. 514–27.

14 Fierer, N., et al., 'Forensic identification using skin bacterial communities.' *Proceedings of the National Academy of Sciences*, 2010. 107(14): pp. 6477–81.

15 Falony, G., et al., 'Population-level analysis of gut microbiome varia-tion.' *Science*, 2016. 352(6285): pp. 560–4.

16 Mellacheruvu, D., et al., 'The CRAPome: a contaminant repository for affinity purification–mass spectrometry data.' *Nature Methods*, 2013. 10(8): pp. 730–6.

17 Salter, S.J., et al., 'Reagent and laboratory contamination can critically impact sequence-based microbiome analyses.' *BMC Biol*, 2014. 12: p. 87.

18 Zhang, S., et al., 'Beer-gut microbiome alliance: a discussion of beer-mediated immunomodulation via the gut microbiome.' *Frontiers in Nutrition*, 2023. 10.

19 Asnicar, F., et al., 'Microbiome connections with host metabolism and habitual diet from 1,098 deeply phenotyped individuals.' *Nature Medicine*, 2021. 27(2): pp. 321–2.

20 Kellogg, J.H., *Plain Facts about Sexual Life*. 1882, London: Ravenswood.

21 Wastyk, H.C., et al., 'Gut-microbiota-targeted diets modulate human immune status.' *Cell*, 2021. 184(16): pp. 4137-53.e14.

22 Paugarten, N., 'Little Helper.' *The New Yorker*, 2003.

23 WHO, *Aspartame hazard and risk assessment results released*. 2023; Available at: https://www.who.int/news/item/14-07-2023-aspartame-hazard-and-risk-assessment-results-released.

24 F.D.A., *Aspartame and Other Sweeteners in Food*. 2023; Available at: https://www.fda.gov/food/food-additives-petitions/aspartame-and-other-sweeteners-food.

25 Del Pozo, S., et al., 'Potential Effects of Sucralose and Saccharin on Gut Microbiota: A Review.' *Nutrients*, 2022. 14(8).

26 Rusch, M., *Sugarless Haribo Gummy Bear Reviews On Amazon Are The Most Insane Thing You'll Read Today*. 2019; Available at: https://www.buzzfeednews.com/article/michaelrusch/haribo-gummy-bear-reviews-on-amazon-are-the-most-insane-thin.

27 Voorhies, A.A., et al., 'Study of the impact of long-duration space missions at the International Space Station on the astronaut microbi-ome.' *Scientific Reports*, 2019. 9(1): p. 9911.

28 Davis-Richardson, A.G., et al., 'Bacteroides dorei dominates gut micro-biome prior to autoimmunity in Finnish children at high risk for type 1 diabetes.' *Frontiers Microbiol*, 2014. 5: p. 678.

29 Asnicar, F., et al., 'Microbiome connections with host metabolism and habitual diet from 1,098 deeply phenotyped individuals.'

30 Abdullah, A., et al., 'The magnitude of association between overweight and obesity and the risk of diabetes: A meta-analysis of prospective cohort studies.' *Diabetes Research and Clinical Practice*, 2010. 89(3): pp. 309–19.

31 Thanarajah, S.E., et al., 'Habitual daily intake of a sweet and fatty snack modulates reward processing in humans.' *Cell Metabolism*, 2023. 35(4): pp. 571–84.e6.

32 Wang, Z., et al., 'IL-1α is required for T cell-driven weight loss after respiratory viral infection.' *Mucosal Immunol*, 2024. 17(2): pp. 272–87.

33 Bailey, C.J., 'Metformin: historical overview.' *Diabetologia*, 2017. 60(9): pp. 1566–76.

34 Gill, S.K., et al., 'Increased airway glucose increases airway bacterial load in hyperglycaemia.' *Sci Rep*, 2016. 6: p. 27636.

35 Slack, C., A. Foley, and L. Partridge, 'Activation of AMPK by the Putative Dietary Restriction Mimetic Metformin Is Insufficient to Extend Lifespan in Drosophila.' *PLoS ONE*, 2012. 7(10): p. e47699.

36 Liao, J., et al., 'Bariatric surgery and health outcomes: An umbrella analysis.' *Frontiers Endocrinology*, 2022. 13: p. 1016613.

37 Castaneda, D., et al., 'Risk of Suicide and Self-harm Is Increased After Bariatric Surgery—a Systematic Review and Meta-analysis.' *Obesity Surgery*, 2019. 29(1): pp. 322–33.

38 Liu, D., et al., 'Calorie Restriction with or without Time-Restricted Eating in Weight Loss.' *New England Journal of Medicine*, 2022. 386(16): pp. 1495–504.

39 Wei, M., et al., 'Fasting-mimicking diet and markers/risk factors for aging, diabetes, cancer, and cardiovascular disease.' *Sci Transl Med*, 2017. 9(377).

40 Pratchett, T. and N. Gaiman, *Good Omens*. 1990, London: Corgi.

41 Robinson, E., 'Veganism and body weight: An N of 1 self-experiment.' *Physiology & Behavior*, 2023. 270: p. 114301.

42 ONS, *Personal well-being user guidance*. 2018; Available from: https://www.ons.gov.uk/peoplepopulationandcommunity/wellbeing/methodologies/personalwellbeingsurveyuserguide.

43 Kroenke, K., R.L. Spitzer, and J.B.W. Williams, 'The PHQ-9.' *Journal of General Internal Medicine*, 2001. 16(9): pp. 606–13.

44 Spitzer, R.L., et al., 'A Brief Measure for Assessing Generalized Anxiety Disorder: The GAD-7.' *Archives of Internal Medicine*, 2006. 166(10): pp. 1092–7.

10 PLUMBING NEW DEPTHS

1 Age UK, *London Loos: The views of older Londoners*. 2022; Available from: https://www.ageuk.org.uk/london/projects-campaigns/out-and-about/london-loos/.

2 Markt, S.C., et al., 'Sniffing out significant "Pee values": genome wide association study of asparagus anosmia.'

3 Eggleton, M.G., 'The diuretic action of alcohol in man.' *J Physiol*, 1942. 101(2): pp. 172–91.

4 Shirreffs, S.M. and R.J. Maughan, 'Restoration of fluid balance after exercise-induced dehydration: effects of alcohol consumption.' *J Appl Physiol* (1985), 1997. 83(4): pp. 1152–8.

5 Desbrow, B., D. Murray, and M. Leveritt, 'Beer as a sports drink? Manipulating beer's ingredients to replace lost fluid.' *Int J Sport Nutr Exerc Metab*, 2013. 23(6): pp. 593–600.

6 Flear, C.T.G., G.V. Gill, and J. Burn, 'Beer Drinking and Hyponatraemia.' *Lancet*, 1981. 318(8244): p. 477.

7 Arca, K.N. and R.B. Halker Singh, 'Dehydration and Headache.' *Curr Pain Headache Rep*, 2021. 25(8): p. 56.

8 Yamada, Y., et al., 'Variation in human water turnover associated with environmental and lifestyle factors.' *Science*, 2022. 378(6622): pp. 909–15.

9 Qian, N., et al., 'Rapid single-particle chemical imaging of nanoplastics by SRS microscopy.' *Proceedings of the National Academy of Sciences*, 2024. 121(3): p. e2300582121.

10 Billingham, R.E., L. Brent, and P.B. Medawar, 'Actively acquired tolerance of foreign cells.' *Nature*, 1953. 172(4379): pp. 603–6.

11 Brett, L., *PMSL*.

12 British Liver Trust, *Liver disease in numbers – key facts and statistics*. 2021; Available from: https://britishlivertrust.org.uk/information-and-support/statistics/.

13 UKHSA, *Hepatitis C prevalence falls by 45% in England*. 2023; Available from: https://www.gov.uk/government/news/hepatitis-c-prevalence-falls-by-45-in-england.

14 Boniface, S. and N. Shelton, 'How is alcohol consumption affected if we account for under-reporting? A hypothetical scenario.' *European Journal of Public Health*, 2013. 23(6): pp. 1076–81.

15 Zhang, S., et al., 'Beer-gut microbiome alliance.'

16 Xi, B., et al., 'Relationship of Alcohol Consumption to All-Cause, Cardiovascular, and Cancer-Related Mortality in U.S. Adults.' *Journal of the American College of Cardiology*, 2017. 70(8): pp. 913–22.

17 Nutt, D., *Drink?: The New Science of Alcohol and Your Health*. 2023, New York: Hachette.

18 Nutt, D., 'Equasy – An overlooked addiction with implications for the current debate on drug harms.' *Journal of Psychopharmacology*, 2009. 23(1): pp. 3–5.

19 Forsyth, A.J.M., 'Distorted? a quantitative exploration of drug fatality reports in the popular press.' *International Journal of Drug Policy*, 2001. 12(5): pp. 435–53.

20 Braidwood, R.J., et al., 'Symposium: Did Man Once Live by Beer Alone?' *American Anthropologist*, 1953. 55(4): pp. 515–26.

21 Liu, L., et al., 'Fermented beverage and food storage in 13,000 y-old stone mortars at Raqefet Cave, Israel: Investigating Natufian ritual feasting.' *Journal of Archaeological Science: Reports*, 2018. 21: pp. 783–93.

22 Kragh, H., 'From Disulfiram to Antabuse: The Invention of a Drug.' *Bulletin for the History of Chemistry*, 2008. 33.

23 Oldham, M., et al., *Youth Drinking in Decline*. 2018, Sheffield: University of Sheffield.

24 Whitaker, V., et al., 'Young people's explanations for the decline in youth drinking in England.' *BMC Public Health*, 2023. 23(1): p. 402.

25 Nutt, D., et al., 'Development of a rational scale to assess the harm of drugs of potential misuse.' *Lancet*, 2007. 369(9566): pp. 1047–53.

26 Barba, T., et al., 'Psychedelics and sexual functioning: a mixed-methods study.' *Scientific Reports*, 2024. 14(1): p. 2181.

11 WHAT A DRAG IT IS

1 McKee, M., et al., 'The changing health needs of the UK population.' *The Lancet*, 2021. 397(10288): pp. 1979–91.

2 Gill, D.M., 'Bacterial toxins: a table of lethal amounts.' *Microbiol Rev*, 1982. 46(1): pp. 86–94.

3 Jones, M.E., et al., 'Osteoarthritis and other long-term health conditions in former elite cricketers.' *Journal of Science and Medicine in Sport*, 2018. 21(6): pp. 558–63.

4 Gui, T., et al., 'Does sports participation (including level of performance and previous injury) increase risk of osteoarthritis? A systematic review and meta-analysis.' *British Journal of Sports Medicine*, 2016. 50(23): p. 1459.

5 Appelboom, T., 'Hypothesis: Rubens—one of the first victims of an epidemic of rheumatoid arthritis that started in the 16th–17th century?' *Rheumatology*, 2005. 44(5): pp. 681–3.

6 NHS, *Finalised Patient Reported Outcome Measures (PROMs) in England for Hip & Knee Replacements, April 2018 – March 2019*. 2020; Available from: https://digital.nhs.uk/data-and-information/publications/statistical/patient-reported-outcome-measures-proms/finalised-hip--knee-replacements-april-2018---march-2019/patient-profile.

7 Brack, A.S. and P. Muñoz-Cánoves, 'The ins and outs of muscle stem cell aging.' *Skeletal Muscle*, 2016. 6(1): p. 1.

8 Elabd, C., et al., 'Oxytocin is an age-specific circulating hormone that is necessary for muscle maintenance and regeneration.' *Nature Communications*, 2014. 5(1): p. 4082.

9 Orkaby, A.R. and A.W. Schwartz, 'Toenails as the "Hemoglobin A1c" of Functional Independence-Beyond the Polished Wingtips.' *JAMA Intern Med*, 2018. 178(5): pp. 598–9.

10 Quinn, T.J., et al., 'Functional assessment in older people.' *BMJ*, 2011. 343: p. d4681.

11 Gawande, A., *Being Mortal: Medicine and What Matters in the End.* 2014, New York: Henry Holt & Company.

12 Witty, C., *Chief Medical Officer's annual report 2023: health in an ageing society*, DHSC, Editor. 2023.

13 Montorsi, F. and M. Oettel, 'Testosterone and sleep-related erections: an overview.' *J Sex Med*, 2005. 2(6): pp. 771–84.

14 Surampudi, P.N., C. Wang, and R. Swerdloff, 'Hypogonadism in the aging male diagnosis, potential benefits, and risks of testosterone replacement therapy.' *Int J Endocrinol*, 2012. 2012: p. 625434.

15 FDA, *Testosterone Products: Drug Safety Communication – FDA Cautions About Using Testosterone Products for Low Testosterone Due to Aging; Requires Labeling Change to Inform of Possible Increased Risk of Heart Attack And Stroke.* 2015; Available from: https://web.archive.org/web/20150305015556/https://www.fda.gov/Safety/MedWatch/SafetyInformation/SafetyAlertsforHumanMedicalProducts/ucm436280.htm.

16 Hales, C., et al., 'Prescription Drug Use Among Adults Aged 40–79 in the United States and Canada.' *NCHS Data Brief*, 2019. 347.

17 Fenerty, S.D., et al., 'The effect of reminder systems on patients' adherence to treatment.' *Patient Prefer Adherence*, 2012. 6: pp. 127–35.

18 Dolgin, E., 'Send in the senolytics.' *Nature Biotechnology*, 2020. 38(12): pp. 1,371–7.

19 Gonzales, M.M., et al., 'Senolytic therapy in mild Alzheimer's disease: a phase 1 feasibility trial.' *Nat Med*, 2023. 29(10): pp. 2,481–8.

20 Raffaele, M. and M. Vinciguerra, 'The costs and benefits of senotherapeutics for human health.' *Lancet Healthy Longevity*, 2022. 3(1): pp. e67–e77.

21 Shin, S., 'Meta-Analysis of the Effect of Yoga Practice on Physical Fitness in the Elderly.' *Int J Environ Res Public Health*, 2021. 18(21).

22 Knapton, S., 'AI-generated nonsense about rat with giant penis published by leading scientific journal', *Telegraph*. 2024.

23 Tregoning, J., 'AI writing tools could hand scientists the "gift of time".'
    Nature, 2023.

12 THINGS SHOULD ONLY GET BETTER

1   Min, K.J., C.K. Lee, and H.N. Park, 'The lifespan of Korean eunuchs.'
    Curr Biol, 2012. 22(18): p. R792-3.
2   Hamilton, J.B. and G.E. Mestler, 'Mortality and Survival:
    Comparison of Eunuchs with Intact Men and Women in a Mentally
    Retarded Population.' *Journal of Gerontology*, 1969. 24(4): pp. 395–411.
3   Le Bourg, É., 'No Ground for Advocating that Korean Eunuchs Lived
    Longer than Intact Men.' *Gerontology*, 2015. 62(1): pp. 69–70.
4   Nieschlag, E., S. Nieschlag, and H.M. Behre, 'Lifespan and testoster-
    one.' *Nature*, 1993. 366(6452): p. 215.
5   Albrecht, P., C. Eimer, and E. Kasten, 'The scrotum: A comparison of
    men's and women's aesthetic assessments.' *Journal of Cosmetic
    Dermatology*, 2023. 22(8): pp. 2273–82.
6   Dattani, S. and L. Rodes-Guirado, 'Why do women live longer than
    men?' *Our World In Data*, 2023.
7   de Magalhaes, J.P., 'Here's to wine, chocolate and a long healthy life.'
    Conversation, 2014.
8   Gupta, V.K., et al., 'Semmelweis Reflex: An Age-Old Prejudice.' *World
    Neurosurg*, 2020. 136: pp. e119–e125.
9   Carolyn, B., 'High price and demand for semaglutide means lack of
    access for US patients.' *BMJ*, 2023. 382: p. p1863.
10 McCartney, G., 'Illustrating health inequalities in Glasgow.' *Journal of
    Epidemiology and Community Health*, 2011. 65(1): p. 94.
11 Chetty, R., et al., 'The Association Between Income and Life Expectancy
    in the United States, 2001–2014.' *JAMA*, 2016. 315(16): pp. 1,750–66.
12 Asaria, M., et al., 'Socioeconomic inequality in life expectancy in India.'
    BMJ Glob Health, 2019. 4(3): p. e001445.
13 Farr, W., 'On the Construction of Life-Tables, Illustrated by a New Life
    -Table of the Healthy Districts of England.' Philosophical Transactions
    of the Royal Society of London, 1859. 149: pp. 837–78.
14 Benzeval, M., et al., *How does money influence health?* 2014. JRF: York.
15 The King's Fund, *New analysis reveals stark inequalities in obesity rates
    across England*. 2021; Available from: https://www.kingsfund.org.uk/
    insight-and-analysis/press-releases/stark-inequalities-obesity-rates-
    across-england.

16 Office for Health Improvement & Disabilities, *Local Health Authority Profiles: Epsom.* 2024; Available from: https://fingertips.phe.org.uk/profile/health-profiles/data#page/1/ati/301/are/E07000208.

17 Najman, J.M., et al., 'The inter- and intra- generational transmission of family poverty and hardship (adversity): A prospective 30 year study.' *PLoS One*, 2018. 13(1): p. e0190504.

18 Mullen, J., 'History of Water Fluoridation.' *British Dental Journal*, 2005. 199(7): pp. 1–4.

19 Milligan, S., *"Rommel?" "Gunner Who?": A Confrontation in the Desert.* 1974, London: Book Club Associates.

20 Remler, D.K., 'Poor smokers, poor quitters, and cigarette tax regressivity.' *Am J Public Health*, 2004. 94(2): pp. 225–9.

21 Smith, J., *Ending smoking could free up 75,000 GP appointments each month.* 2023; Available at: https://news.cancerresearchuk.org/2023/03/07/ending-smoking-could-free-up-gp-appointments/.

22 Nina Trivedy, R., et al., 'Estimated impact of the UK soft drinks industry levy on childhood hospital admissions for carious tooth extractions: interrupted time series analysis.' *BMJ Nutrition, Prevention & Health*, 2023. 6(2): p. 243.

23 Rogers, N.T., et al., 'Associations between trajectories of obesity prevalence in English primary school children and the UK soft drinks industry levy: An interrupted time series analysis of surveillance data.' *PLOS Medicine*, 2023. 20(1): p. e1004160.

24 Goodier, M., 'Fire crews in England deal with obesity callouts every four hours.' *Guardian*, 2023.

25 Lean, M.E., et al., 'Primary care-led weight management for remission of type 2 diabetes (DiRECT): an open-label, cluster-randomised trial.' *Lancet*, 2018. 391(10120): pp. 541–51.

26 Dimbleby, H., *National Food Strategy*, DEFRA, Editor. 2021.

27 Berman, M.G., J. Jonides, and S. Kaplan, 'The cognitive benefits of interacting with nature.' *Psychol Sci*, 2008. 19(12): pp. 1207–12.

28 Ulrich, R.S., 'View Through a Window May Influence Recovery from Surgery.' *Science*, 1984. 224(4647): p. 420–41.

29 Furuyashiki, A., et al., 'A comparative study of the physiological and psychological effects of forest bathing (Shinrin-yoku) on working age people with and without depressive tendencies.' *Environ Health Prev Med*, 2019. 24(1): p. 46.

CONCLUSION: LIVE FOREVER OR AGE BETTER?

1 Watson, J.L., et al., 'Macromolecular condensation buffers intracellular water potential.' *Nature*, 2023. 623(7988): pp. 842–52.

2 Light, N., et al., 'Knowledge overconfidence is associated with anti-consensus views on controversial scientific issues.' *Science Advances*, 2022. 8(29): p. eabo0038.

3 Baumeister, R.F., et al., 'Ego depletion: is the active self a limited resource?' *J Pers Soc Psychol*, 1998. 74(5): pp. 252–65.

4 Hagger, M.S., et al., 'A Multilab Preregistered Replication of the Ego-Depletion Effect.' *Perspectives on Psychological Science*, 2016. 11(4): pp. 546–73.

5 Job, V., et al., 'Beliefs about willpower determine the impact of glucose on self-control.' *Proceedings of the National Academy of Sciences*, 2013. 110(37): pp. 14837–42.

6 Vaillant, G.E., *Triumphs of Experience: The Men of the Harvard Grant Study*. 2012, Cambridge, Mass.: Belknap Press.

7 Veenhoven, R., 'Healthy happiness: effects of happiness on physical health and the consequences for preventive health care.' *Journal of Happiness Studies*, 2008. 9(3): pp. 449–69.

8 Lawrence, E.M., R.G. Rogers, and T. Wadsworth, 'Happiness and longevity in the United States.' *Social Science & Medicine*, 2015. 145: pp. 115–9.

9 Danner, D.D., D.A. Snowdon, and W.V. Friesen, 'Positive emotions in early life and longevity: findings from the nun study.' *J Pers Soc Psychol*, 2001. 80(5): pp. 804–13.

10 Maruta, T., et al., 'Optimists vs Pessimists: Survival Rate Among Medical Patients Over a 30-Year Period.' *Mayo Clinic Proceedings*, 2000. 75(2): pp. 140–3.

11 Office of the Surgeon, G., *Publications and Reports of the Surgeon General, in Our Epidemic of Loneliness and Isolation: The U.S. Surgeon General's Advisory on the Healing Effects of Social Connection and Community*. 2023, Washington, D.C.: US Department of Health and Human Services.

12 Holt-Lunstad, J., T.F. Robles, and D.A. Sbarra, 'Advancing social connection as a public health priority in the United States.' *Am Psychol*, 2017. 72(6): pp. 517–30.

13 Kretzler, B., H.H. König, and A. Hajek, 'Pet ownership, loneliness, and social isolation: a systematic review.' *Soc Psychiatry Psychiatr Epidemiol*, 2022. 57(10): pp. 1,935–57.

14 Sidik, S.M., 'Why loneliness is bad for your health.' *Nature*, 2024. 628(8006): pp. 22–4.

15 Doane, L.D. and E.K. Adam, 'Loneliness and cortisol: momentary, day-to-day, and trait associations.' *Psychoneuroendocrinology*, 2010. 35(3): pp. 430–41.

16 Steptoe, A., J. Wardle, and M. Marmot, 'Positive affect and health-related neuroendocrine, cardiovascular, and inflammatory processes.' *Proceedings of the National Academy of Sciences*, 2005. 102(18): pp. 6508–12.

17 Gawande, A., *2021 Stanford Commencement address by Dr. Atul Gawande*. 2021; Available at: https://news.stanford.edu/stories/2021/06/2021-commencement-address-dr-atul-gawande.

18 Poulain, M., et al., 'Identification of a geographic area characterized by extreme longevity in the Sardinia island: the AKEA study.' *Experimental Gerontology*, 2004. 39(9): pp. 1423–9.

19 Buettner, D., *The Secrets of Long Life*. 2005, National Geographic.

20 Franco, M., et al., 'Impact of Energy Intake, Physical Activity, and Population-wide Weight Loss on Cardiovascular Disease and Diabetes Mortality in Cuba, 1980–2005.' *American Journal of Epidemiology*, 2007. 166(12): pp. 1374–80.

21 De Rooij, S.R., et al., 'Lessons learned from 25 Years of Research into Long term Consequences of Prenatal Exposure to the Dutch famine 1944–45: The Dutch famine Birth Cohort.' *International Journal of Environmental Health Research*, 2022. 32(7): pp. 1432–46.

22 Snowdon, D.A., et al., 'Linguistic ability in early life and cognitive function and Alzheimer's disease in late life. Findings from the Nun Study.' *JAMA*, 1996. 275(7): pp. 528–32.

CODA: SUNLIT UPLANDS

1 Gawande, A., *Being Mortal*.

2 Thompson, H.S. and D. Brinkley, *The Proud Highway: Saga of a Desperate Southern Gentleman*, 1955-1967. 1997, New York: Villard.

Index

Locators in *italics* refer to figures and those in **bold** to tables.